FOUNDATION
GCSE MATHEMATICS
FOR WJEC

**Wyn Brice, Linda Mason,
Tony Timbrell**

HOMEWORK BOOK

Hodder Murray
A MEMBER OF THE HODDER HEADLINE GROUP

Hodder Headline's policy is to use papers that are natural, renewable and recyclable products and made from wood grown in sustainable forests. The logging and manufacturing processes are expected to conform to the environmental regulations of the country of origin.

Orders: please contact Bookpoint Ltd, 130 Milton Park, Abingdon, Oxon OX14 4SB. Telephone: (44) 01235 827720. Fax: (44) 01235 400454. Lines are open from 9 a.m. to 5 p.m., Monday to Saturday, with a 24-hour message-answering service. Visit our website at www.hoddereducation.co.uk.

© Wyn Brice, Linda Mason, Tony Timbrell, 2006
First published in 2006 by
Hodder Murray, an imprint of Hodder Education,
a member of the Hodder Headline Group,
338 Euston Road,
London, NW1 3BH

Impression number   10 9 8 7 6 5 4 3 2
Year   2011 2010 2009 2008 2007

Cover photo © Garry Gay/Photographer's Choice/Getty Images
Typeset in 10/12 Times by Tech-Set Ltd, Gateshead, Tyne & Wear
Printed in Great Britain by Martins The Printers, Berwick-upon-Tweed.

A catalogue record for this title is available from the British Library

ISBN: 978 0340 900 178

CONTENTS

INTRODUCTION

This book contains exercises designed to be used for the Foundation tier of GCSE Mathematics. It is particularly aimed at WJEC's Linear Specification and each exercise matches one in the WJEC Foundation Student's Book.

In the Homework Book, the corresponding exercises have the same number and end in H. Thus, for example, if you have been working on Interpreting graphs in class and used Exercise 22.1, then the homework exercise is 22.1H. The homework exercises cover the same mathematics. Some chapters also have exercises corresponding to the review exercises in the textbook.

Some questions are intended to be completed without a calculator, just as in the textbook. These are shown with a non-calculator icon in the same way. Doing these questions without a calculator is vital preparation for the non-calculator section of the examination papers.

These homework exercises provide extra practice and are also in a smaller book to carry home! If you have understood the topics, you should be able to tackle these exercises confidently as they are no harder than those you have done in class and in some cases may be a little easier. See if you agree.

You will find the answers to this homework book in the Foundation Assessment Pack.

1 ➔ INTEGERS, POWERS AND ROOTS 1

EXERCISE 1.1H

1 Work out these.
   (a) 46 + 32   (b) 68 + 35
   (c) 86 + 59   (d) 178 + 26
   (e) 185 + 232   (f) 188 + 346

2 Work out these.
   (a) 87 − 42   (b) 72 − 26
   (c) 72 − 39   (d) 239 − 87
   (e) 273 − 148   (f) 634 − 276

3 Work out these.
   (a) 32 × 4   (b) 28 × 6
   (c) 38 × 3   (d) 49 × 8
   (e) 135 × 7   (f) 278 × 5

4 Work out these.
   (a) 69 ÷ 3   (b) 54 ÷ 3
   (c) 96 ÷ 6   (d) 92 ÷ 4
   (e) 165 ÷ 5   (f) 232 ÷ 4

5 Meena bought a top for £24, a pair of jeans for £37 and a belt for £9. What was the total cost?

6 Karim has £50.
He bought two CDs for £13 each. How much did he have left?

7 Liz bought eight pens at 38p each. What was the total cost in pence?

8 Will organises a disco. The cost of hiring the hall and the disc jockey is £468. How many £6 tickets does he need to sell to cover the cost?

EXERCISE 1.2H

1 List the following.
   (a) The multiples of 9 less than 100
   (b) The multiples of 12 less than 100

2 Use your answers to question **1** to list the common multiples of 9 and 12 less than 100.

3 Look at these numbers.

   4, 7, 9, 16, 18, 21, 25, 30, 42.

   (a) Which have 2 as a factor?
   (b) Which have 3 as a factor?
   (c) Which have 7 as a factor?

4 List the following.
   (a) The multiples of 20 less than 130
   (b) The multiples of 25 less than 130

5 Use your answers to question **4** to find a common multiple of 20 and 25 less than 130.

6 List the following.
   (a) The factors of 20
   (b) The factors of 36

7 Use your answers to question **6** to list the common factors of 20 and 36.

8 List the following.
   (a) The factors of 60
   (b) The factors of 24

9 Use your answers to question **8** to list the common factors of 60 and 24.

10 Round these numbers to the nearest 1000.
(a) 32 300   (b) 203 900
(c) 6600   (d) 243 497
(e) 3 503 776

11 Round these numbers to the nearest 100.
(a) 8392   (b) 21 830
(c) 354   (d) 756 982
(e) 7032

12 Here are some newspaper headlines. Round the numbers so that they have more impact.
(a) Blues win with a majority of 7832!
(b) £2 127 836 wasted by 'red tape'!
(c) Moody goes to Real for €34 632 578!
(d) Number passing goes up by 12 364!

EXERCISE 1.3H

1 Work out these.
(a) 71×10 (b) 84×100
(c) 26×1000 (d) 402×100
(e) $78 \times 10\,000$ (f) 80×100
(g) 617×1000 (h) 2800×100
(i) 140×1000 (j) $84 \times 10\,000$
(k) $974 \times 100\,000$ (l) $76 \times 1\,000\,000$

2 Work out these.
(a) $590 \div 10$
(b) $29\,000 \div 100$
(c) $648\,000 \div 1000$
(d) $92\,000 \div 100$
(e) $9\,200\,000 \div 1000$
(f) $789\,000 \div 10$
(g) $458\,200 \div 100$
(h) $8\,400\,000 \div 100$
(i) $71\,000\,000 \div 10\,000$

3 Work out these.
(a) 40×30 (b) 60×90
(c) 80×300 (d) 400×400
(e) 700×50 (f) 60×50
(g) 800×4000 (h) 500×200
(i) 900×8000 (j) 6000×2000
(k) $60\,000 \times 30$ (l) 7000×9000

4 Work out these.
(a) 53×20 (b) 63×40
(c) 165×50 (d) 73×400
(e) 82×700 (f) 59×600
(g) 537×800 (h) 96×3000

5 Work out these.
(a) 72×24 (b) 64×36
(c) 71×96 (d) 59×23
(e) 88×39 (f) 252×47
(g) 348×65 (h) 546×83
(i) 792×96 (j) 85×384

6 (a) How many centimetre are there in 589 metres?
(b) Change 4900 centimetres into metres.

7 1 hectare is 10 000 square metres. How many square metres is 63 hectares?

8 Jeans cost £30 per pair. What will eight pairs cost?

9 A sales representative drives 400 miles per working day. How far does he drive in a year if he works on 235 days?

10 138 people attended a dinner dance. If the tickets were £26 each, what was the total amount they paid?

EXERCISE 1.4H

1 Work out these without your calculator.
    (a) 1^2 **(b)** 8^2 **(c)** 2^4
    (d) 40^2 **(e)** 70^2 **(f)** 80^2
    (g) 300^2 **(h)** 500^2 **(i)** 600^2
    (j) 110^2

2 Use your calculator to work out these.
    (a) 18^2 **(b)** 34^2 **(c)** 52^2
    (d) 78^2 **(e)** 86^2 **(f)** 47^2
    (g) 57^2 **(h)** 240^2 **(i)** 389^2
    (j) 643^2

3 Use your calculator to work out these.
    (a) 7^3 **(b)** 8^3
    (c) 12^3 **(d)** 18^3
    (e) 34^3 **(f)** 48^3
    (g) 56^3 **(h)** 231^3

4 Use your calculator to work out these.
    (a) $\sqrt{256}$ **(b)** $\sqrt{441}$
    (c) $\sqrt{576}$ **(d)** $\sqrt{1024}$
    (e) $\sqrt{3969}$ **(f)** $\sqrt{5476}$
    (g) $\sqrt{3364}$ **(h)** $\sqrt{7921}$

5 Work out these without your calculator.
    (a) $\sqrt{1600}$ **(b)** $\sqrt{4900}$ **(c)** $\sqrt{8100}$
    (d) $\sqrt{10\,000}$ **(e)** $\sqrt{90\,000}$

EXERCISE 1.5H

1 Work out these.
    (a) -3 add 5
    (b) -8 add 4
    (c) 7 subtract 11

2 The temperature is $-7°C$.
    Find the new temperature after
    (a) a rise of 4°C.
    (b) a rise of 10°C.
    (c) a fall of 6°C.

3 Find the difference in temperature
    between
    (a) 8°C and 26°C.
    (b) $-3°C$ and 12°C.
    (c) $-23°C$ and $-11°C$.

4 Arrange these numbers in order,
    smallest first.
    (a) $2, -6, 5, -4$
    (b) $7, -5, 0, -8$
    (c) $-2, 6, -7, -8, 9, 3$

5 An aircraft flies at 15 000 m where the
    outside temperature is $-70°C$.
    When the aircraft lands the outside
    temperature is 21°C.
    What is the difference between these
    temperatures?

EXERCISE 2.1H

1 Match each of the following angles to its diagram.

295°   45°   168°   25°   90°

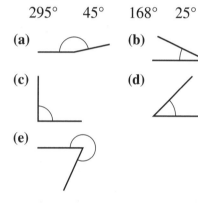

(a)   **(b)**

(c)   **(d)**

(e)

2 Are these angles acute, right angle, obtuse or reflex?

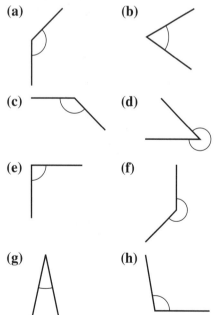

(a)   **(b)**

(c)   **(d)**

(e)   **(f)**

(g)   **(h)**

3 Are angles of these sizes acute, right angle, obtuse or reflex?
   (a) 57°   **(b)** 110°
   (c) 90°   **(d)** 195°
   (e) 28°   **(f)** 124°
   (g) 345°   **(h)** 91°
   (i) 222°   **(j)** 6°

EXERCISE 2.2H

Work out the size of the unknown angle in each of these diagrams.

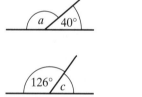

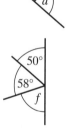

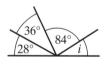

EXERCISE 2.3H

Work out the size of the unknown angle in each of these diagrams.

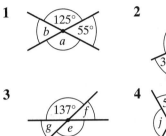

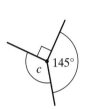

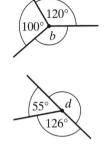

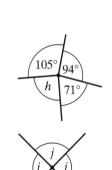

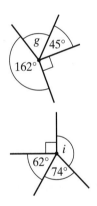

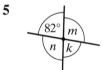

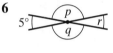

EXERCISE 2.4H

For each question
- make a copy of the diagram.
- work out the size of each unknown angle.
- give a reason for each answer.

1   2

3   4

5   6

7   8

9   10

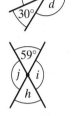

REVIEW EXERCISE 2H

For each question, first make a copy of the diagram.

1

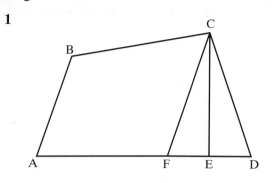

Using the above diagram
(a) name the line which is perpendicular to AD
(b) name the line which is parallel to AB.

WJEC Summer 2002

2

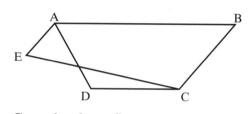

Copy the above figure,
(i) name the line which is parallel to DC
(ii) draw a line through A which is perpendicular to the line EC.

WJEC Summer 2003

3

Copy the above diagram.
(i) Draw a line through the point D, which is perpendicular to AB.
(ii) Draw a line through the point C, which is parallel to AB.

WJEC Summer 2004

4

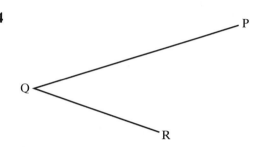

Copy the above diagram.
(i) Draw a line through the point R which is parallel to the line QP.
(ii) Draw a line through the point P which is perpendicular to the line QP.

WJEC Summer 2005

EXERCISE 3.1H

1 (a) There are 6 men and 4 women on a bus. How many people are there on the bus?
   (b) There are 6 men and w women on a bus.
       Write an expression for the number of people on the bus.
   (c) There are m men and w women on a bus.
       Write an expression for the number of people on the bus.

2 (a) Shamsah buys 4 apples and 3 oranges.
       How many pieces of fruit does she buy?
   (b) Shamsah buys x apples and 3 oranges.
       Write an expression for the number of pieces of fruit she buys altogether.
   (c) Shamsah buys x apples and y oranges.
       Write an expression for the number of pieces of fruit she buys altogether.

3 (a) What is the length of this line?
       4 ————— 7 - - - - -
   (b) Write an expression for the length of this line.
       x ————— 7 - - - - -

(c) Write an expression for the length of this line.
    4 ————— x - - - - -

4 (a) What is the length of this line?
       3 ————— 5 - - - - -
   (b) Write an expression for the length of this line.
       p ————— 5 - - - - -
   (c) Write an expression for the length of this line.
       3 ————— q - - - - -

5 (a) Write an expression for the length of this line.
       x ————— x - - - -
   (b) Write an expression for the length of this line.
       2 ——— 4 - - - x ———
   (c) Write an expression for the length of this line.
       x ——— 4 - - - y ———

6 (a) Write an expression for the length of this line.
       x ——— y - - - 2 ———
   (b) Write an expression for the length of this line.
       x ——— 5 - - - x ———

(c) Write an expression for the length of this line.

$$x \quad y \quad x$$

7 The length of a rectangle is 6 cm longer than its width.
   (a) What is the length of the rectangle if the width is
       (i) 12 cm?   (ii) 16 cm?
   (b) Write an expression for the length of the rectangle if the width is w cm.

8 There are x people on a bus. At a bus stop, 5 more get on.
   Write an expression for the number of people now on the bus.

9 Cox apples cost 10p more a kilogram than Braeburn apples.
   (a) What is the cost of a kilogram of Cox apples when a kilogram of Braeburn apples costs
       (i) 85 pence?   (ii) 73 pence?
   (b) Write an expression for the cost of a kilogram of Cox apples when a kilogram of Braeburn apples costs x pence.

10 At Pete's chip shop, chips cost c pence a bag.
    Write an expression for the cost of
    (a) 2 bags.   (b) 4 bags.   (c) 8 bags.

EXERCISE 3.2H

1 Pam has 7 fewer books than James.
   (a) How many books does Pam have if James has
       (i) 12 books?   (ii) 20 books?
   (b) Write an expression for the number of books Pam has if James has b books.

2 Tea cost 20p a cup less than coffee.
   (a) What is the cost of a cup of tea if a cup of coffee costs
       (i) 70p?   (ii) 95p?
   (b) Write an expression for the cost of a cup of tea if a cup of coffee costs c pence.

3 Write an expression for the length of the dashed part of these lines.

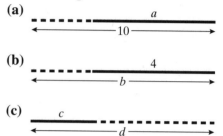

4 The width of a rectangle is 6 cm less than the length.
   (a) What is the width of the rectangle if the length is
       (i) 12 cm?   (ii) 16 cm?
   (b) Write an expression for the width of the rectangle if the length is y cm.

5 Mike earns £50 less per week than Adrian.
   (a) How much does Mike earn per week if Adrian earns
       (i) £160?   (ii) £340?
   (b) Write an expression for how much Mike earns per week if Adrian earns £P.

6 There were 24 people on a bus when it arrived at a stop. Some got off.
   (a) How many people were left on the bus if the number getting off was
       (i) 2?   (ii) 5?

(b) Write an expression for the number of people left on the bus if the number getting off was s.

7 Joshua bought a coat that cost £x less than the coat Jeremy bought.
Write an expression for the cost of Josh's coat if Jeremy's cost
(a) £65. **(b)** £80. **(c)** £y.

8 The width of a rectangle is half its length.
(a) What is the width of the rectangle if the length is
(i) 12 cm? **(ii)** 16 cm?
(b) Write an expression for the width of the rectangle if the length is y cm.

9 (a) Write an expression for the length of the dashed part of these lines.
(i)
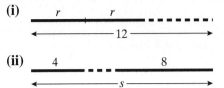
(ii)
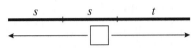

(b) Write an expression for the length of this line.

10 Joe spends £x in d days. He spends the same amount each day.
Write an expression for the amount he spends in one day.

4 → FRACTIONS

EXERCISE 4.1H

1 What fraction is
   (a) 4 of 12?   (b) 18 of 24?
   (c) 9 of 15?   (d) 14 of 21?
   (e) 18 of 36?   (f) £20 of £80?
   (g) 7 cm of 56 cm?   (h) 22 g of 77 g?
   Write the fractions in their lowest terms.

2 Copy and complete the following.

   (a) $\dfrac{1}{3} = \dfrac{\square}{6} = \dfrac{3}{\square} = \dfrac{10}{12} = \dfrac{\square}{300}$

   (b) $\dfrac{1}{7} = \dfrac{2}{\square} = \dfrac{\square}{21} = \dfrac{4}{70} = \dfrac{100}{\square}$

3 Copy and complete the following

   (a) $\dfrac{3}{5} = \dfrac{\square}{15}$ (b) $\dfrac{10}{18} = \dfrac{5}{\square}$

   (c) $\dfrac{1}{2} = \dfrac{\square}{24}$ (d) $\dfrac{50}{70} = \dfrac{5}{\square}$

   (e) $\dfrac{12}{22} = \dfrac{\square}{11}$ (f) $\dfrac{4}{7} = \dfrac{12}{\square}$

   (g) $\dfrac{5}{11} = \dfrac{\square}{66}$ (h) $\dfrac{3}{27} = \dfrac{1}{\square}$

   (i) $\dfrac{2}{9} = \dfrac{\square}{36}$ (j) $\dfrac{4}{13} = \dfrac{\square}{39}$

   (k) $\dfrac{25}{45} = \dfrac{5}{\square}$ (l) $\dfrac{63}{70} = \dfrac{\square}{10}$

4 Express these fractions in their lowest terms
   (a) $\frac{10}{14}$ (b) $\frac{3}{18}$ (c) $\frac{16}{20}$ (d) $\frac{18}{30}$
   (e) $\frac{21}{35}$ (f) $\frac{30}{50}$ (g) $\frac{24}{36}$ (h) $\frac{32}{56}$
   (i) $\frac{4}{52}$ (j) $\frac{70}{140}$ (k) $\frac{20}{160}$ (l) $\frac{200}{1000}$
   (m) $\frac{18}{54}$ (n) $\frac{42}{56}$ (o) $\frac{60}{72}$ (p) $\frac{75}{125}$

EXERCISE 4.2H

1 Find $\frac{1}{3}$ of these quantities.
   (a) 30   (b) 36   (c) 69
   (d) £90   (e) £15

2 Find $\frac{1}{5}$ of these quantities.
   (a) 35   (b) 55   (c) 90
   (d) £140   (e) 36 m

3 Find $\frac{3}{4}$ of these quantities.
   (a) 36   (b) 60   (c) 100
   (d) £68   (e) £180

4 Find $\frac{5}{6}$ of these quantities.
   (a) 60   (b) 36   (c) 90
   (d) 48 cm   (e) £150

5 Faheem receives £24 for his birthday.
   He saves $\frac{1}{4}$ of it.
   How much does he save?

6 Each week Abigail earns £20.
   She spends $\frac{1}{5}$ of it on sweets and drinks
   and saves $\frac{3}{10}$ of it.
   (a) How much does she spend on
       sweet and drinks?
   (b) How much does she save?

7 The maths budget is £800.
$\frac{3}{5}$ is spent on text books.
How much is left?

8 There are 180 students in year 9.
$\frac{4}{9}$ of them come to school by bus.
How many come by bus?

9 A school has 720 students.
$\frac{5}{8}$ of them have school dinners.
How many students have school
dinners?

10 Which is the larger, a $\frac{3}{8}$ share of £160
or a $\frac{4}{5}$ share of £80?
Show your working.

EXERCISE 4.3H

1 Work out these. Where possible cancel
the fractions to their lowest terms
(a) $\frac{1}{6} \times 5$ (b) $\frac{1}{9} \times 4$ (c) $\frac{2}{5} \times 2$
(d) $3 \times \frac{2}{9}$ (e) $4 \times \frac{1}{12}$

2 Change these improper fractions to
mixed numbers.
(a) $\frac{11}{4}$ (b) $\frac{12}{5}$ (c) $\frac{14}{3}$ (d) $\frac{21}{2}$
(e) $\frac{23}{4}$ (f) $\frac{23}{6}$ (g) $\frac{33}{8}$ (h) $\frac{12}{11}$
(i) $\frac{19}{4}$ (j) $\frac{79}{10}$

3 Work out these.
Write your answers first as improper
fractions, then as mixed numbers.
Where possible cancel the fractions to
their lowest terms
(a) $\frac{1}{2} \times 11$ (b) $\frac{3}{5} \times 4$ (c) $\frac{3}{7} \times 4$
(d) $6 \times \frac{5}{7}$ (e) $4 \times \frac{4}{9}$ (f) $\frac{5}{8} \times 4$
(g) $\frac{3}{10} \times 6$ (h) $10 \times \frac{3}{8}$ (i) $7 \times \frac{3}{5}$
(j) $\frac{5}{11} \times 7$ (k) $\frac{5}{6} \times 8$ (l) $9 \times \frac{9}{10}$
(m) $10 \times \frac{5}{8}$ (n) $\frac{5}{12} \times 8$ (o) $7 \times \frac{5}{21}$

EXERCISE 5.1H

Simplify these expressions.

1 $a + a + a$
2 $b + b + b + b + b + b + b + b$
3 $c \times 6$
4 $2 \times d$
5 $5 \times c + 2 \times c$
6 $c + c + c + c - c + c$
7 $2a + 3a - 2a$
8 $5 \times b$
9 $3c - 2c$
10 $4a + 3a$
11 $6x + 4x - 2x$
12 $7a - a$
13 $4y - 3y$
14 $2b + 4b - b$
15 $2p - 3p + 4p$
16 $6s + 2s - 3s - 4s$
17 $b + 2b$
18 $a + 2a - 3a$
19 $3 \times b + 4 \times b$
20 $5a + 2a$

EXERCISE 5.2H

Simplify these expressions.

1 $4a + 2b + a + b$
2 $3a + 2b - 2a$
3 $5 \times c \times d$
4 $5c + 3d$
5 $a + 2b + 4a + b$
6 $2 \times p + 3 \times q$
7 $3 \times a \times 4 \times b$
8 $2x + 4y + y + 2x$
9 $5 \times a \times a$
10 $6a + b + 3c + b + 3a + 3c$
11 $4a + 2a - 3a + a$
12 $2 \times a \times b + 4a \times a$
13 $3a - 2b$
14 $b \times b$
15 $4s + 2 - 3s + 3$
16 $a + 2b - a + 2b$
17 $2 \times a \times b \times b$
18 $3x + 2y + 4 + 2y - 2x + 3$
19 $a + 2b + 2a - 3b$
20 $2a \times 2b \times 3a \times b$

EXERCISE 6.1H

1 Calculate the third angle in each of these triangles.

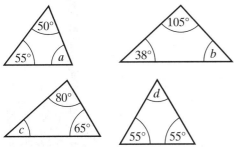

2 PQR is a triangle. Angle QPR = 48° and angle PQR = 60°.
Sketch the triangle and find angle PRQ.

3 PQR is a triangle. PRS is a straight line. Angle QPR = 48° and angle PQR = 63°.

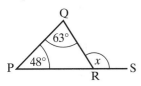

Calculate angle x.

4 ABCD is a rectangle.
P is halfway between B and C.

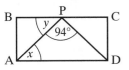

(a) What type of triangle is APD?
(b) Calculate the angles x and y.

5 In this diagram LM and NP are straight lines.

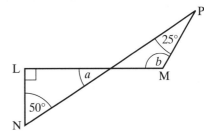

Calculate the size of angles a and b.

EXERCISE 6.2H

1 Which of these triangles are a pair of congruent triangles?

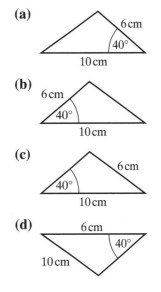

2 Which of these triangles are congruent to triangle A?

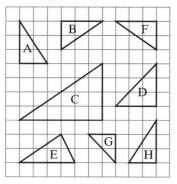

3 These two triangles are congruent. What is the size of angle x?

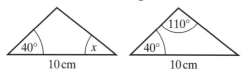

4 Two triangles are congruent.
One triangle has two angles of 70°.
Write down the size of all three angles in the other triangle.

5 Draw a rectangle ABCD.
Draw the line AC. This divides the rectangle into two congruent triangles.
Label your diagram clearly to show which angles are equal.

6 Complete these statements.
(a) In two congruent triangles, corresponding angles are ………
(b) In two congruent triangles, corresponding sides are ………

EXERCISE 6.3H

1 Look at these shapes.

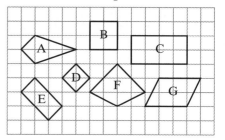

Copy and complete this table. Put a tick (✔) if the description fits. Put a cross (✗) if the description does not fit.

Shape	All sides equal	Opposite sides parallel	All angles right angles
A			
B			
C			
D			
E			
F			
G			

2 Name each of the shapes in question **1**.

3 (a) What is the name of a quadrilateral that has two pairs of equal sides which are next to each other?
(b) What other fact is true about this quadrilateral?

4 Sketch a trapezium. Label clearly any equal sides and any parallel sides.

5 A quadrilateral has opposite sides equal and parallel.
Name all the different shapes this quadrilateral could be.

EXERCISE 6.4H

1 Here are some shapes made out of centimetre squares.

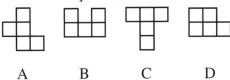

A	B	C	D

Which ones will not fold up to make an open box?

2 A box has edges of length 5 cm, 6 cm and 3 cm.
Is it a cube or a cuboid? Explain your answer.

3 Draw a net of this closed box.
Label the length of each edge.

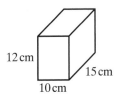

EXERCISE 6.5H

These solid shapes are made from centimetre cubes.
Use isometric paper to draw the shapes.

1   **2**

3   **4**

5    **6**

EXERCISE 7.1H

1 Which of the following are discrete data and which are continuous data?

   Time of day
   Weight of an egg
   Number of peas in a pod
   Day you were born
   Amount of rain in a week
   Price of an item

2 List at least three more examples of
   (a) continuous data.
   (b) discrete data.

3 Design a data collection sheet for each of the following.
   (a) Favourite school subject
   (b) Favourite sport
   (c) School shoe colour

EXERCISE 7.2H

1 Draw a vertical line graph to show each of these sets of data.

(a)

Type of fast food	Frequency
Pizza	18
Chinese	8
Burger	13
Kebab	6
Fish and chips	4
Other	25

(b)

Type of vehicle	Frequency
Car	43
Van	29
Lorry	15
Bus	7
Motorbike	4
Other	2

(c)

Type of tree	Frequency
Oak	3
Beech	7
Pine	10
Sycamore	6
Lime	2
Other	2

2 Draw a bar chart to show each of these sets of data.

(a)

Favourite colour	Frequency
Blue	23
Red	9
Green	6
Yellow	4
Pink	7
Black	3
Other	8

(b)

Number of sisters	Frequency
0	7
1	12
2	8
3	2
4 or more	1

(c)

Shoe size	Frequency
39	12
40	8
41	15
42	11
43	9
44	6
45	2
Other	7

(d)

Type of bird	Frequency
Sparrow	41
Starling	32
Pigeon	14
Finch	5
Dove	2
Other	6

(e)

Drinks per day	Frequency
Less than 4	5
4	12
5	21
6	28
7	23
8 or more	11

(f)

Number of pets	Frequency
0	14
1	45
2	23
3	12
4	5
5 or more	1

(g) Sally does a survey to find out what pupils think of school lunch. Her results are shown in the following table.

Satisfaction level	Replies	Frequency
Excellent	卌 卌 卌 卌 卌 II	
Good	卌 卌 卌 III	
Fair	卌 卌 卌 I	
Poor	卌 卌 卌 卌 卌 II	

Complete the frequency column in the above table and use this data to draw a suitable bar chart.

WJEC Summer 2001

(h) Ann carried out a survey of the colour of cars parked at a hotel near her home. She recorded the colours in the following way.

**R – red   B – blue   G – green
W – white   S – silver
O – other colours**

The results of her survey are as follows.

**R R B W W S O O O R S
S S W G G R R B S S O
O W R S S G B G B W O
S S R R W R S G G S W**

(i) Using the above data complete the following.

Colour	Replies	Frequency
Red	⋕ ||||	9
Blue		
Green		
White		
Silver		
Other colours		

(ii) Copy and complete the above table and use this data to draw a suitable bar chart.

WJEC Summer 2004

(i) Tom carries out a survey, in his class, to find the colour of pupils eyes. His results are shown in the following table.

Colour	Number of pupils	Frequency
Blue	⋕ ⋕ |||	
Green	⋕ ||	
Brown	⋕ ⋕ ⋕	

Complete the frequency column in the above table and use this data to draw a suitable bar chart.

WJEC Summer 2005

EXERCISE 7.3H

1 Here is a two-way table showing the results of a car survey.
Copy and complete the table.

	Black	Not black	Total
German	18	49	
Not German	23	64	
Total			

2 Here is a two-way table showing the results of a survey of vehicles in a car park.
(a) Copy and complete the table.

	White	Not white	Total
Car	3	86	
Not car	24	7	
Total			

(b) How many white cars were in the survey?
(c) How many cars were in the survey?
(d) How many vehicles were not white?
(e) How many vehicles were surveyed in total?

3 Here is a two-way table showing the sales made by a bookshop on one day.
   (a) Copy and complete the table.

	Paperback	Not paperback	Total
Fiction		11	98
Non-fiction	9	17	
Total			

   (b) How many non-fiction books were sold?
   (c) How many paperback fiction books were sold?
   (d) How many books were sold in total?

4 A group of students were questioned about their favourite subject and their preferred writing hand.
   Here is a two-way table showing the results.
   (a) Copy and complete the table.

	Right hand	Left hand	Neither	Total
Maths	27	8	1	
Not maths		13		112
Total	126			

   (b) How many students were left-handed?
   (c) How many students chose maths?
   (d) How many right-handed students chose a subject other than maths?
   (e) How many students were in the survey?

5 At a swimming championship, Australia, China and the USA won most medals. Their results are shown in the table.

   (a) Copy and complete the table.

	Gold	Silver	Bronze	Total
Australia		27	13	74
China		20		63
USA	21	19		59
Total	71			

   (b) How many silver medals were awarded in total to these countries?
   (c) How many gold medals did Australia win?
   (d) Which of these countries won most bronze medals?
   (e) How many medals were awarded in total to these countries?

EXERCISE 7.4H

1 Draw a frequency table using tallies for each of these sets of data.
   (a) Marks out of 30 in a maths test
      Use groups of 1 to 5, 6 to 10, 11 to 15, 16 to 20, ...

   14 26 16 24 13   4   8 22 20 24
    2 25 15 16 22 11 10 23   8 23
   23 17   5   7 17   9 12 18 17 18
    7 29   8 18 14 16 13 16   6   6
   13 16 14   9   6 19   7 13 15   4
   19 21 19 14   9   3   7 23 19 21

   (b) Number of apples on a tree
      Use groups of 1 to 10, 11 to 20, 21 to 30, ...

   43 34 17 37 18 42 11 37 18 15
   24 25 23 23 30 24 13 16 28 21
   15 37 35 27 28 33 21 27   6 18
   23   2 29 39 31 40 36 86 27 46
    7 25 21 19 90 29 27 24 14 27
   36 21 30 25 22 31 29 37   7 43

(c) Number of lorries delivering to a depot per day
Use groups of 0 to 19, 20 to 39, 40 to 59, ...

```
49  41  63  36  29  53  65  46  27  48
51  74  52  44  47  77  87  35  69  74
53  27  44  50  67   9  36  14  31  42
73  55  35  53  83  45  62  53  43  61
 4  62  28  37  16  43  45  32  55  43
66  37  17  18  33  38  57  51  77  66
```

2 Draw a bar chart for each of these sets of data.

(a)

Number of video games	Frequency
1 to 5	7
6 to 10	14
11 to 15	31
16 to 20	27
21 to 25	11
26 or more	10

(b)

Number of swimmers	Frequency
0 to 49	2
50 to 99	5
100 to 149	11
150 to 199	9
200 to 249	3
250 or more	1

(c)

Number of eggs	Frequency
1 to 5	5
6 to 10	8
11 to 15	17
16 to 20	23
21 to 25	9
26 or more	3

(d)

Number of runners	Frequency
1 to 5	9
6 to 10	17
11 to 15	18
16 to 20	4
21 or more	2

(e)

Number of balloons	Frequency
0 to 9	4
10 to 19	11
20 to 29	27
30 to 39	35
40 to 49	21
50 to 59	13
60 to 69	5
70 or more	9

(f)

Number of fish	Frequency
1 to 10	9
11 to 20	9
21 to 30	15
31 to 40	28
41 to 50	7
51 or more	2

EXERCISE 8.1H

1 The total number of seats in an
assembly hall is found by multiplying
the number of rows by 20.
How many seats are there when there
are
(a) 20 rows?   (b) 15 rows?
(c) 17 rows?   (d) $13\frac{1}{2}$ rows?

2 The perimeter of a square is found
by multiplying the length of one side
by 4.
Work out the perimeter of a square
with sides of these lengths.
(a) 5 cm   (b) 13 cm
(c) $6\frac{1}{2}$ cm   (d) 8.2 cm

3 The cost of a child's bus fare is half the
cost of an adult's fare. What is the cost
of a child's fare when the adult fare is
(a) 80p?   (b) 70p?
(c) £1.30?   (d) £2.36?

4 Five friends divide a bag of sweets
equally.To work out how many sweets
each person receives, divide the total
number of sweets by 5.
How many sweets does each person
receive when the bag contains
(a) 20 sweets?   (b) 55 sweets?
(c) 80 sweets?   (d) 125 sweets?

5 The time, in minutes, needed to cook a
piece of beef can be found by
multiplying the weight of the beef in
kilograms by 40. How long does it
take to cook a piece of beef weighing
(a) 2 kg?   (b) 8 kg?
(c) $5\frac{1}{2}$ kg?   (d) 3.2 kg?

6 The area of a rectangular room is
found by multiplying the length by the
width. Work out the area of these
rooms.
(a) Length 4 m and width 3 m
(b) Length and width both 7 m
(c) Length $2\frac{1}{2}$ m and width 6 m
(d) Length 5.4 m and width 5 m

7 A car hire firm charges a fee of £30,
plus £2 for each mile travelled.
How much is the total charge if you
travel
(a) 25 miles?   (b) 70 miles?
(c) 400 miles?   (d) 185 miles?

8 A football team earns 3 points for
every match they win and 1 point for
every match they draw. How many
points in total do they earn when they
(a) win 3 games and draw 5?
(b) win 7 games and draw 10?
(c) win 14 games and draw 6?
(d) win 28 games and draw 3?

9 To change American dollars into pounds, divide the number of dollars by 1.5.
How many pounds you will get for
(a) $6?   (b) $150?
(c) $24?   (d) $10.50?

10 To change a distance from miles into kilometres, multiply the number of miles by 8 and divide by 5.
Work out the number of kilometres that is the same as
(a) 10 miles.   (b) 15 miles.
(c) 100 miles.   (d) 44 miles.

EXERCISE 8.2H

In questions **1** to **7**, write down a formula for each situation using the letters in **bold**.

1 The total number of **seats** in an assembly hall is found by multiplying the number of **rows** by 20.

2 The **perimeter** of a square is found by multiplying the **length** of one side by 4.

3 The cost of a **child**'s bus fare is half the cost of an **adult**'s fare.

4 Five friends divide a bag of sweets equally.
To work out how many sweets **each** person receives, divide the total number of **sweets** by 5.

5 The **time**, in minutes, needed to cook a piece of beef can be found by multiplying the **weight** of the beef in kilograms by 40.

6 The **area** of a rectangular room is found by multiplying the **length** by the **width**.

7 A car hire firm **charges** a fee of £30, plus £2 for each **mile** travelled.

In questions **8** to **10**, write down a formula for the situation using appropriate letters and say what each letter stands for.

8 A football team earns 3 points for every match they win and 1 point for every match they draw.

9 To change American dollars into pounds, divide the number of dollars by 1.5.

10 To change a distance from miles into kilometres, multiply the number of miles by 8 and divide by 5.

EXERCISE 8.3H

1 Find the value of these expressions when $a = 4$ and $b = 2$.
(a) $a + b$ (b) $a - b$
(c) $2a$ (d) $7a$
(e) $5b$ (f) $8b$
(g) ab (h) a^2
(i) $3ab$ (j) $2a + b$
(k) $3a - b$ (l) $5a + 4b$
(m) $3a - 4b$ (n) $a^2 + b^2$
(o) $3a^2$ (p) $6b^2$
(q) a^2b (r) b^3
(s) $\dfrac{a}{b}$ (t) $\dfrac{10b}{a}$
(u) $b - a$ (v) $\dfrac{4b}{2a}$
(w) $\dfrac{a^2}{b^2}$ (x) $\dfrac{b}{a}$

2 Use the formula $C = 5t + 2$ to find C when
(a) $t = 2$.   (b) $t = 8$.   (c) $t = 12$.
(d) $t = 20$.   (e) $t = \frac{1}{2}$.

3 Use the formula $P = 2C - F$ to find P when
(a) $C = 4, F = 4$.
(b) $C = 3, F = 5$.
(c) $C = 8, F = 5$.
(d) $C = 0, F = 7$.

4 Use the formula $X = w + nd$ to find X when
(a) $w = 1, n = 2, d = 1$.
(b) $w = 4, n = 2, d = 6$.
(c) $w = 12, n = 10, d = 0$.
(d) $w = 50, n = 30, d = 10$.
(e) $w = 11, n = 9, d = \frac{1}{2}$.

5 Find the value of these expressions when $k = 3$ and $m = 2$.
(a) $k + m$ (b) $5k$ (c) km
(d) $m - k$ (e) $4k + 6m$

REVIEW EXERCISE 8H

1 Use the formula:

total cost = number of days worked $\times$ £2000 + £500

Find the **total cost** when the **number of days worked** = 6.
WJEC Summer 2005

2 The following formula is used to calculate the cost of repairing a gas boiler:

cost = rate per hour $\times$ number of hours + standing charge

(a) Calculate the **cost** when:

rate per hour = £4.00
number of hours = 5
standing charge = £25.00

(b) Calculate the **standing charge** when:

cost = £50.00
rate per hour = £6.00
number of hours = 3
WJEC Summer 2000

3 The cost, in pounds, of hiring a van is given by the following formula:

cost = number of hours $\times$ 5 + 20

(a) Find the cost of hiring a van for 6 hours
(b) Tim paid £85. For how many hours did Tim hire the van?
WJEC Summer 2001

4 The cost, in pounds, of hiring a cement mixer is given by the formula:

cost = number of hours $\times$ 7 + 20

(a) Find the cost when the cement mixer is hired for 12 hours.
(b) Find the **number of hours** a cement mixer is hired when the cost is £125.
WJEC Summer 2002

5 The formula shown below is used to calculate the amount paid, in pounds, to repair a television set.

Amount paid = 12.25 $\times$ number of hours + 55

(a) Calculate the amount paid when the number of hours is 13.
(b) Calculate the **number of hours** when the amount paid is 153 pounds.
WJEC Summer 2004

EXERCISE 9.1H

1 Write in words the place value of the digit 8 in each of these numbers.
   (a) 800   (b) 0.8   (c) 8000
   (d) 8.74   (e) 0.028

2 Write these numbers as decimals.
   (a) $\frac{9}{10}$ (b) $8\frac{3}{10}$ (c) $\frac{7}{100}$
   (d) $61\frac{23}{100}$ (e) $\frac{183}{1000}$

3 Write these decimals as fractions or mixed numbers (whole numbers and fractions) in their lowest terms.
   (a) 0.4   (b) 5.1   (c) 27.5
   (d) 0.25   (e) 1.13

4 Write all these amounts in pounds. Then arrange them in order, smallest first.
   £13.84 82p £4.09 £0.08 12p £5.79

5 Write these numbers in order, largest first.
   0.816   8.61   0.068   1.806   60.1

6 Use the fact that there are 1000 millilitres in a litre to write these capacities in litres.
   (a) 250 ml   (b) 1500 ml
   (c) 185 ml   (d) 5 ml
   (e) 2250 ml

7 Write these lengths in centimetres.
   (a) 1.2 m   (b) 84 mm
   (c) 3.85 m   (d) 56 mm
   (e) 7 mm

8 Write these lengths in order, smallest first.
   7.3 cm 0.61 m 83 mm 2.08 m 48.1 cm

9 Draw a grid to show that $\frac{2}{5} = \frac{4}{10} = 0.4$.

10 Write these fractions as decimals.
   (a) $\frac{7}{10}$ (b) $\frac{1}{2}$ (c) $\frac{4}{5}$
   (d) $\frac{3}{4}$ (e) $\frac{13}{100}$

EXERCISE 9.2H

1 Work these out.

   (a) $\begin{array}{r} 3.87 \\ +9.15 \\ \hline \end{array}$ (b) $\begin{array}{r} 38.43 \\ +59.12 \\ \hline \end{array}$

   (c) $\begin{array}{r} 41.53 \\ +67.42 \\ \hline \end{array}$ (d) $\begin{array}{r} 16.19 \\ +\ 8.34 \\ \hline \end{array}$

   (e) $\begin{array}{r} 36.84 \\ +53.92 \\ \hline \end{array}$ (f) $\begin{array}{r} 137.58 \\ +273.16 \\ \hline \end{array}$

2 Work these out.

   (a) $\begin{array}{r} 19.28 \\ -\ 6.25 \\ \hline \end{array}$ (b) $\begin{array}{r} 47.16 \\ -15.42 \\ \hline \end{array}$

   (c) $\begin{array}{r} 253.80 \\ -\ 81.47 \\ \hline \end{array}$ (d) $\begin{array}{r} 17.84 \\ -\ 8.13 \\ \hline \end{array}$

   (e) $\begin{array}{r} 57.42 \\ -23.51 \\ \hline \end{array}$ (f) $\begin{array}{r} 721.58 \\ -236.82 \\ \hline \end{array}$

3 Work out these.
(a) £3.95 + 82p + £1.57
(b) £15.21 + 77p + £3.42 + 61p
(c) £63.84 + 90p +£8.51 +
£91.20 + 47p
(d) £6.25 + 42p + 87p + £63.20

4 Find the cost of six DVDs at £13.45 each.

5 In the javelin, Amina throws 52.15 m and Raj throws 46.47 m.
Find the difference between the lengths of their throws.

6 The times for the first and last places in a show-jumping event were 47.82 seconds and 53.19 seconds.
Find the difference between these times.

7 Melissa buys three of these bags of carrots.

(a) What is the total weight?
(b) What is the total cost?

8 Find the cost of 3 kg of onions at £0.72 per kilogram.

9 Work out these.
(a) 7.3 m + 87 cm + 6.1 m
(b) 5.2 m + 54 cm + 3.85 m + 76 cm
(c) 6.1 m − 278 cm
(d) 4.5 m − 57 cm
(e) 12.1 cm − 7 mm

10 Work out these.
(a) 650 g + 1.7 kg + 56 g + 2.1 kg
(b) 3.7 kg + 450 g − 1.5 kg
(c) 1.4 kg − 586 g
(d) 2 litres − 730 ml
(e) 3 × 0.58 litres

11 Holly buys two scarves at £8.45 each and a pair of shoes at £35.99.
How much change does she get from £60?

12 Dean buys two newspapers at 55p each and three magazines at £1.20 each.
How much change does he get from £10?

13 Work out these.
(a) 6 × 0.4   (b) 0.2 × 7
(c) 5 × 0.3   (d) 0.8 × 9
(e) 0.4 × 0.1   (f) 0.7 × 0.8
(g) 50 × 0.7   (h) 0.3 × 80
(i) 0.4 × 0.3   (j) 0.6 × 0.1
(k) $(0.7)^2$ (l) $(0.2)^2$

10 → CIRCLES AND POLYGONS

EXERCISE 10.1H

1 Name these parts of a circle.

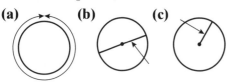

(a)   (b)   (c)

2 Name these polygons.

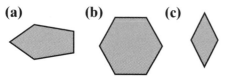

(a)   (b)   (c)

3 A regular hexagon is constructed in a circle.
How many degrees are measured at the centre to draw each radius required?

4 Draw a circle of radius 5 cm and use it to construct a regular pentagon.
Measure the length of a side of your pentagon.

5 Draw a circle of radius 6 cm and use it to construct a regular nine-sided polygon.
Measure the length of a side of your polygon.

6

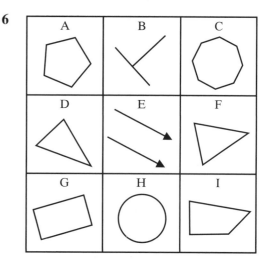

Copy and complete the following table to show which of the above boxes contains:

a circle	H
two parallel lines	
an isosceles triangle	
a pentagon	
a rectangle	

WJEC Summer 2001

7   In each of parts **(i)**, **(ii)** and **(iii)** a
    diagram of one of the named shapes
    has been drawn. Copy these diagrams
    and draw the other named shape,
    showing clearly the difference
    between the two shapes.

(i)   A square and a rectangle.

(ii)   An isosceles triangle and an
           equilateral triangle.

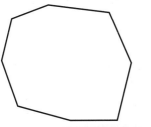

(iii) An octogon and a pentagon.

WJEC Summer 2005

8   Copy the following diagrams and draw
    lines to show

    an arc of a circle
    a radius of a circle
    a tangent of a circle
    a diameter of a circle
    a chord of a circle

    An arc of a circle has been draw for
    you.

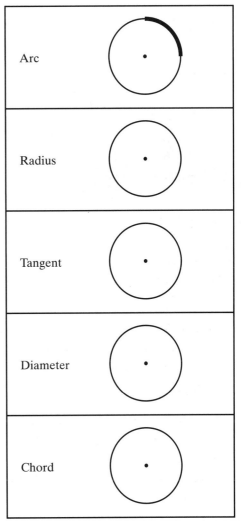

WJEC Summer 2005

EXERCISE 11.1H

Solve these equations.

1 $x - 5 = 12$ 2 $5x = 40$
3 $9y = 45$ 4 $x + 7 = 18$
5 $10p = 60$ 6 $2x = 2$
7 $3x = 18$ 8 $6y = 48$
9 $x - 2 = -5$ 10 $x - 12 = -4$
11 $x - 10 = -2$ 12 $x - 11 = -11$

EXERCISE 11.2H

Solve these equations.

1 $15 - x = 9$ 2 $7 - x = 10$
3 $3x + 2 = 11$ 4 $4x + 1 = 17$
5 $3x - 5 = 7$ 6 $3x + 6 = 48$
7 $3x - 7 = 41$ 8 $8x - 3 = 29$
9 $4x + 12 = 20$ 10 $4x + 18 = 22$
11 $5x + 6 = 31$ 12 $2x - 4 = -2$

EXERCISE 11.3H

Solve these equations.

1 $\dfrac{x}{4} = 7$ 2 $\dfrac{a}{3} = 6$ 3 $\dfrac{p}{7} = 7$

4 $\dfrac{y}{4} = 12$ 5 $\dfrac{x}{2} = 18$ 6 $\dfrac{x}{4} = 6$

7 $\dfrac{x}{3} = 7$ 8 $\dfrac{x}{5} = 5$ 9 $\dfrac{x}{6} = 1$

10 $\dfrac{x}{3} = 3$ 11 $\dfrac{x}{3} = 9$ 12 $\dfrac{x}{4} = 20$

EXERCISE 11.4H

1 Robbie thinks of a number.
   He adds 3 to it. The answer is 8.
   Use x to represent Robbie's number.
   Write down and solve an equation to
   find Robbie's number.

2 Patsy thinks of a number.
   She multiplies it by 3. The answer is 15.
   Use x to represent Patsy's number.
   Write down and solve an equation to
   find Patsy's number.

3 Four chocolate biscuits cost 64p.
   Use c to represent the cost of one
   biscuit in pence.
   Write down and solve an equation to
   find the cost of one biscuit.

4 Patel has 16 CDs. Sonia has x CD's.
   Together they have 34 CDs.
   Write down and solve an equation to
   find how many CDs Sonia has.

5 Joy had x pens.
   She gives away 6 of them and has 7 left.
   Write down and solve an equation to
   find how many pens Joy had to start
   with.

EXERCISE 12.1H

1 The number of cars of different colours passing a school one lunchtime were as follows.

Black   15
Blue   25
Green   10
Red   20
Silver   30
Other   35

Draw a pictogram to show this data using ● to represent 10 cars.

2 The number of fish caught each day one week in a fishing lake was as follows.

Monday   12
Tuesday   8
Wednesday 10
Thursday   4
Friday   9
Saturday   18
Sunday   27

Draw a pictogram to show this data using ■ to represent four fish.

3 The number of people taking an early morning swim each day one week at a leisure centre was as follows.

Monday   16
Tuesday   14
Wednesday 18
Thursday   15
Friday   16
Saturday   10
Sunday   8

Create your own symbol for four swimmers and draw a pictogram to show this data.

4 The number of umbrellas sold each month one year by a department store was as follows.

January   40
February   20
March   60
April   80
May   50
June   30
July   15
August   10
September 25
October   55
November   40
December   65

Draw a pictogram to show this data.

EXERCISE 12.2H

1 Draw a pie chart for each of these sets of data.

(a)

Favourite drink	Frequency
Soft drink	42
Milk	12
Water	48
Juice	18
Other	60
Total	180

(b)

Eye colour	Frequency
Blue	76
Blue/Green	28
Brown	92
Grey	38
Other	6
Total	240

(c)

Type of programme	Frequency
Comedy	14
Soap	22
Cartoon	16
Drama	18
Other	10
Total	80

2 Draw a pie chart for each of these sets of data.

(a)

Activity	Hours
School	6
Sleeping	9
Eating	2
Playing	3
TV	2
Other	2

(b)

Favourite TV channel	Frequency
BBC	12
ITV	15
Channel 4	6
5	9
Satellite/Cable	30

(c)

Favourite type of film	Frequency
Comedy	25
Horror	14
Romance	32
Action	17
Other	2

EXERCISE 12.3H

1 The table shows the maximum daytime temperature in Miami over a period of 12 days.

Day	Mon	Tues	Wed	Thur	Fri	Sat	Sun	Mon	Tues	Wed	Thur	Fri
Temp. (°C)	27	29	28	31	33	34	37	36	33	32	30	31

Draw a line graph to show this information.

2 The table shows the monthly sales of gravel by a garden centre in 2005.

Month	Jan	Feb	Mar	Apr	May	June	July	Aug	Sept	Oct	Nov	Dec
Sales (tonnes)	2	3	5	8	13	11	6	5	7	8	4	1

Draw a line graph to show this information.

3 The table shows the number of tickets sold by a cinema one week.

Day	Mon	Tues	Wed	Thur	Fri	Sat	Sun
Tickets	58	71	49	86	183	205	152

Draw a line graph to show this information.

4 The table shows the monthly sales of widgets in 2005.

Month	Jan	Feb	Mar	Apr	May	June	July	Aug	Sept	Oct	Nov	Dec
Sales (×£1000)	3	5	8	13	24	36	29	18	13	9	6	4

Draw a line graph to show this information.

EXERCISE 13.1H

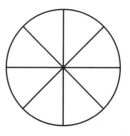

1 Change these percentages to fractions. Write your answers in their lowest terms.
   (a) 45%   (b) 85%
   (c) 4%   (d) 130%

2 Change these percentages to decimals.
   (a) 18%   (b) 26%   (c) 92%
   (d) 8%   (e) 42%   (f) 3%
   (g) 12%   (h) 1%   (i) 170%
   (j) 350%   (k) 5%   (l) 14.5%

3 Change these decimals to percentages.
   (a) 0.73   (b) 0.28   (c) 0.06
   (d) 0.235   (e) 1.68

4 Change these fractions to percentages.
   (a) $\frac{3}{10}$ (b) $\frac{4}{5}$ (c) $\frac{13}{20}$
   (d) $\frac{7}{25}$ (e) $\frac{18}{40}$

5 Write $\frac{1}{4}$ as a decimal   ....
   Write 0.5 as a fraction   ....
   Write 30% as a decimal ....
   Write $\frac{1}{4}$, 0.5 and 30% in descending order.

6 Write $\frac{2}{5}$ as a decimal   ....
   Write 0.25 as a fraction ....
   Write 10% as a decimal ....
   Write $\frac{2}{5}$, 0.25 and 10% in order of size, starting with the smallest first.
   WJEC Summer 2003

7 (a) Write $\frac{1}{2}$ as a decimal
       Write 0.4 as a fraction
       Write 60% as a decimal
       Write $\frac{1}{2}$, 0.4 and 60% in ascending order.
   (b) Shade 75% of the following shape.

   WJEC Summer 2002

8 Showing all your working, find which of the quantities 0.8, $\frac{17}{20}$ and 84% is (i) the smallest, (ii) the largest.
   WJEC Summer 2004

9 Showing all your working, find which of the quantities $\frac{9}{20}$, 40% and 0.42 is (a) the smallest, (b) the largest.
   WJEC Summer 2005

EXERCISE 13.2H

Do not use your calculator for questions 1 to 6.

1 Find 25% of £84.

2 Find 60% of 35 kg.

3 Find 15% of £70.

4 James borrowed £80 000 to buy a house and paid 7% interest in the first year. Calculate the interest.

5 30% of the students in Year 7 came from one primary school.
There are 220 students in Year 7.
How many came from that primary school?

6 15% of Mr and Mrs Thompson's Council Tax is used to pay for police, fire service and civil defence. Mr and Mrs Thompson's Council tax bill is £1200.
How much goes to pay for police, fire service and civil defence?

You may use your calculator for questions **7** to **12**.

7 Find 19% of £36.

8 Find 37% of 240 metres.

9 Find 108% of £64.

10 92% of the seats at a concert were sold out after 1 week.
There were 32 000 seats available.
How many were sold out in the first week?

11 Sarah pays 6% of her earnings into a pension fund.
She earns £1450 per month
How much does she pay into the pension fund each month?

12 Michael changes £350 into euros. The bank charges him 2.5% commission.
How much is the commission?

EXERCISE 13.3H

Do not use your calculator for questions **1** to **6**.

1 Increase £700 by these percentages.
   (a) 40%   **(b)** 35%
   (c) 70%   **(d)** 21%

2 Decrease £520 by these percentages.
   (a) 70%   **(b)** 25%
   (c) 6%   **(d)** 20%

3 Michelle earns £16 000 per year.
She receives a salary increase of 3%.
Find her new salary.

4 A company cuts its wage bill by 15%.
Its wage bill before the cut was £4 600 000.
What was the wage bill after the cut?

5 Amy earned £360 per week.
She pays 11% national insurance and 15% income tax.
Calculate Amy's weekly take-home pay.

6 A furniture shop has a sale.

40% off

Complete the table to find the sale price of these articles.

Item	Original Price (£)	Reduction (£)	Sale price (£)
Bookcase	180		
Dining table and chairs	840		
Sofa	440		
Display unit	260		

You may use your calculator for questions 7 to 12.

7 Increase £84 by these percentages.
   (a) 14%   (b) 36%
   (c) 9%   (d) 72%

8 Decrease £428 by these percentages.
   (a) 23%   (b) 37%
   (c) 7%   (d) 69%

9 The value of a car fell by 12% in the first year.
   It cost £16 800 when new.
   What was its value after 1 year?

10 8% more people passed an examination in 2005 than in 2004.
   6400 passed in 2004.
   How many passed in 2005?

11 In 1971 there were 902 000 births in the UK. By 1981 the birth rate had fallen by 19%.
   Calculate the number of births in 1981. Give your answer to the nearest thousand.

12 An antique increased in value by 180% in five years. It was worth £240 at the start of the five years.
   What was it worth at the end of the five years?

EXERCISE 13.4H

Do not use your calculator for questions 1 to 5.

1 Calculate £4 as a percentage of £50.

2 Calculate 8 metres as a percentage of 40 metres.

3 Calculate 90p as a percentage of £3.

4 Graeme earns £50 per week. He receives a pay increase of £2 per week.
   Calculate his pay increase as a percentage of £50.

5 A computer originally costing £600 is reduced by £90.
   Calculate the reduction as a percentage of the original price.

You may use your calculator for questions 6 to 10.

6 Calculate £64 as a percentage of £400.

7 Calculate 180 kg as a percentage of 400 kg.

8 The number of houses built in a town in 2003 was 480. This increased by 168 in 2004.
   Find the increase as a percentage of 480.

9 The value of an investment fell from £4800 in 1998 to £4200 in 2003.
   Find the reduction as a percentage of £4800.

10 In 2005 the national minimum wage was raised from £4.85 to £5.05.
   Calculate the increase as a percentage of £4.85.
   Give your answer to the nearest whole number.

14 → COORDINATES

EXERCISE 14.1H

1 Write down the coordinates of the points A, B, C, D, E, F, G, H, I and J.

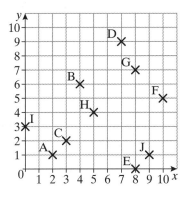

2 On a grid, draw *x*- and *y*-axes from 0 to 8. Plot and label these points.
A(7, 1) B(3, 5) C(4, 2)
D(2, 0) E(0, 6)

EXERCISE 14.2H

1 Write down the coordinates of the points A, B, C, D, E, F, G, H, I and J.

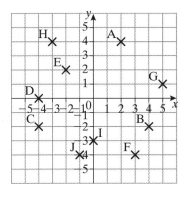

In questions **2** to **9** you will need to draw *x*- and *y*-axes from −5 to 5.

2 Plot and label the points A(5, 1), B(5, −3), C(−3, −3) and D(−3, 1). Join the points to make the shape ABCD. What is the special name of the shape ABCD?

3 Plot and label the points A(4, 1), B(4, −4) and C(−3, −4). Join the points to make the shape ABC. What is the special name of the shape ABC?

4 Plot and label the points A(2, 4), B(2, −2), C(−1, −4) and D(−1, 2). Join the points to make the shape ABCD. What is the special name of the shape ABCD?

5 Plot and label the points A(2, 2), B(5, −3), C(−1, −3) and D(−2, 2). Join the points to make the shape ABCD. What is the special name of the shape ABCD?

6 Plot and label the points A(1, 3), B(1, −4) and C(−3, −4). Mark the point D so that ABCD is a rectangle. Write down the coordinates of D.

7 Plot and label the points A(1, 1), B(5, −1) and D(−3, −1). Mark the point C so that ABCD is a rhombus. Write down the coordinates of C.

8 Plot and label the points A(−1, 4), B(1, 0) and C(−3, −2). Mark the point D so that ABCD is a square. Write down the coordinates of D.

9 Plot and label the points A(−2, 2), B(3, 3) and C(1, −2). Mark the point D so that ABCD is a parallelogram. Write down the coordinates of D.

EXERCISE 14.3H

1 Write down the equation of each of the lines (a), (b), (c) and (d).

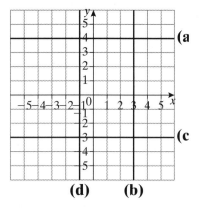

(d)   (b)

For questions **2** to **4** use a single grid with axes labelled from −7 to 7.

2 Complete this table of values for the equation $y = x + 2$.

x	−3	−2	−1	0	1	2	3
y = x + 2	−1		1		3		

Plot the points on your grid and join them with a straight line.

3 Complete this table of values for the equation $y = x − 4$.

x	−3	−2	−1	0	1	2	3	4	5
y = x − 4	−7		−5		−3		−1		

Plot the points on your grid and join them with a straight line.

4 Complete this table of values for the equation $x + y = 5$.

x	−1	0	1	2	3	4	5	6
y = x + 2	6		4		2		0	

Plot the points on your grid and join them with a straight line.

5 What are the coordinates of the point where the lines in questions **3** and **4** cross?

For questions **6** and **7** use a single grid with axes labelled from −8 to 8.

6 Complete this table of values for the equation $y = 2x − 3$.

x	−2	−1	0	1	2	3	4	5
2x		−2		2			8	
−3		−3		−3			−3	
y = 2x − 3		−5		−1			5	

Plot the points on your grid and join them with a straight line.

7 Complete this table of values for the equation $y = 6 − x$. Remember: $6 − (−2) = 6 + 2 = 8$.

x	−2	−1	0	1	2	3	4	5	6	7
y = 6 − x	8		5			2			0	

Plot the points on your grid and join them with a straight line.

8 What are the coordinates of the point where the lines in questions **6** and **7** cross?

EXERCISE 15.1H

1 Copy these shapes.
   On each shape, draw the lines of
   symmetry.

   (a)   **(b)**   **(c)**

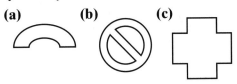

2 How many lines of symmetry do these
   arrows have?

   (a)   **(b)**   **(c)**

3 Draw a rectangle.
   A rectangle has two lines of symmetry.
   Draw both lines of symmetry on your
   rectangle.

4 This triangle has three lines of
   symmetry.
   What type of triangle is it?

5 Copy this grid.
   Shade more squares so that the diagonal
   broken line is the line of symmetry.

6 Copy this grid.
   Shade more squares so that the grid
   has two lines of symmetry.

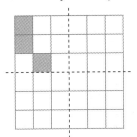

7 Make a pattern
   with two lines
   of symmetry.
   Shade squares on
   a grid like this.

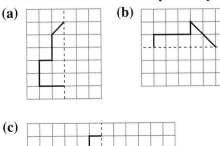

8 Copy these diagrams.
   Complete the diagrams so that the
   broken lines are lines of symmetry.

   (a)   **(b)**

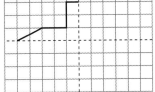

   (c)

EXERCISE 15.2H

1 What is the order of rotation symmetry for each of these shapes?

(a) 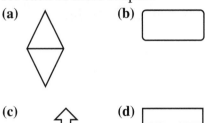   (b)

(c) 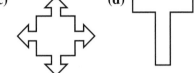   (d)

2 Look at this regular pentagon.

How many lines of symmetry does it have?
What is its order of rotation symmetry?

3 This triangle has rotation symmetry of order 3.

Draw a different triangle that has no rotation symmetry.

4 Copy this grid.
Shade more squares so that the pattern has rotation symmetry of order 2.

5 Copy this grid.
Shade more squares so that the pattern has rotation symmetry of order 4.

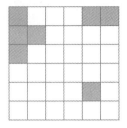

6 Shade squares on a 3 × 3 grid so that your pattern has rotational symmetry of order 2 but no reflection symmetry.

7 Copy this diagram.

Complete it so that it has rotation symmetry of order 4.

8 Copy this diagram.

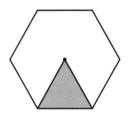

Complete it so that it has rotation symmetry of order 3, but no reflection symmetry.

9 Draw a simple pattern that has rotation symmetry of order 2.

EXERCISE 15.3H

1 Copy each shape on to squared paper
   and reflect it in the mirror line shown.

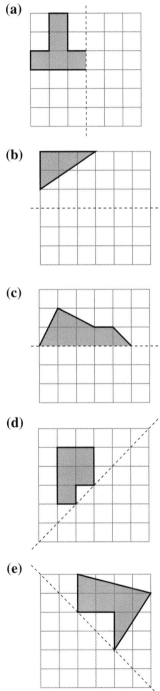

(a)

(b)

(c)

(d)

(e)

2 Copy each pair of shapes on to
   squared paper and draw the mirror line
   for the reflection.

(a)

(b)

(c)

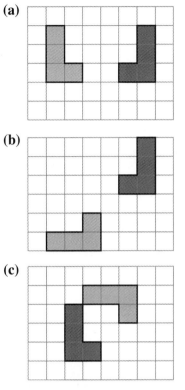

3 Copy the diagram.
   Reflect flag A in the mirror line. Label
   the image B.
   Reflect flag A in the y-axis. Label the
   image C.

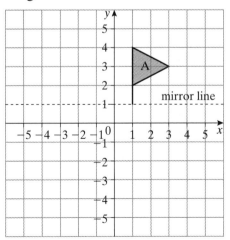

4 Draw x- and y-axes from −5 to 5.
Plot the points (2, 1), (5, 1), (5, 3)
and (3, 5).
Join the points to form a trapezium.
Label it A.
Reflect shape A in the y-axis. Label
the image B.
Reflect shape A in the x-axis. Label
the image C.

5 Through what angle is the dark object
rotated to fit the light image?

(a)

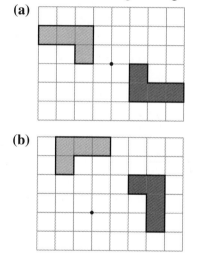

(b)

6 Points A, B and C are part of this shape.
It has rotation symmetry of order 6.

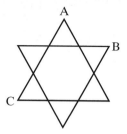

What clockwise angle of rotation maps
(a) A on to B?
(b) A on to C?

7 Triangle A has been rotated or
reflected to give these image triangles.

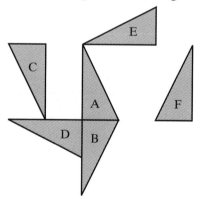

(a) List the triangles which are
reflections of triangle A.
(b) List the triangles which are
rotations of triangle A.

EXERCISE 16.1H

1 Find the median of each of these sets of data.
   (a) 1 4 5 7 9 9 10
   (b) 3 9 6 4 3 9 3 2 7
   (c) 4 5 9 6 9 9 7 4 8 8
   (d) 2 5 3 2 7 8 9 3 4
   (e) 3 7 1 3 5 7 9 11

2 Find the median of these seven weekly wages.
   £257   £238   £289   £362
   £321   £411   £306

3 The ages of six teachers are as follows.
   46   52   38   25   31   62
   Find the median age

4 These are Sophie's marks in five tests.
   5 8 9 7 8
   These are Carole's marks in six tests.
   7 6 9 4 8 3
   (a) Find the median mark for
        (i) Sophie.   **(ii)** Carole.
   (b) Who had the higher median mark?

5 Stephen is asked to find the median of these numbers.
   4 2 1 5 7
   He says the median is 1. He is wrong.
   (a) What has he done wrong?
   (b) What is the correct answer?

EXERCISE 16.2H

1 Find the mode of this set of data.
   1   2   1   3   2   4
   3   5   2   5   2

2 The number of boys who were at ten different sports practices were as follows.
   18   17   17   20   16
   20   9   9   17   10
   Find the modal number of boys at a practice.

3 The marks scored in a test were as follows.
   10   12   18   17   16   18   14   15
   16   18   14   18   20   9   18   13
   Find the modal number of marks scored.

4 Here is a list of the heights of ten people.
   173 cm   158 cm   161 cm   163 cm
   181 cm   153 cm   173 cm   170 cm
   162 cm   180 cm
   Find the modal height.

5 The ages of a group of friends are as follows.
   18   22   21   19   18
   18   17   19   17   21
   Find the modal age.

EXERCISE 16.3H

1 Find the mean of this set of data.

   3   6   5   4   7   6   8   1

2 A gardener measures the heights of a group of plants.
   The heights were as follows.

   50 cm   65 cm   80 cm
   40 cm   35 cm

   Find the mean height.

3 In a test the marks were as follows.

   9   7   8   7   5   6   8   9   5   6

   Find the mean mark.

4 The ages of a group of friends are as follows.

   18   22   21   19   18
   18   17   19   17   21

   Find the mean age.

5 The salaries of ten workers in a small company are as follows.

   £20 000   £20 000
   £20 000   £18 000
   £22 000   £23 000
   £25 000   £21 000
   £23 000   £28 000

   Find the mean salary.

EXERCISE 16.4H

1 Find the range of this set of data.

   3   6   5   4   7   6   8   1

2 A gardener measures the heights of a group of plants.
   The heights were as follows.

   50 cm   65 cm   80 cm
   40 cm   35 cm

   Find the range of the heights.

3 In a test the marks were as follows.

   9   7   8   7   5   6   8   9   5   6

   Find the range of the marks.

4 The ages of a group of friends are as follows.

   18   22   21   19   18
   18   17   19   17   21

   Find the range of their ages.

5 The salaries of ten workers in a small company are as follows.

   £20 000   £20 000
   £20 000   £18 000
   £22 000   £23 000
   £25 000   £21 000
   £23 000   £28 000

   Find the range of their salaries.

EXERCISE 16.5H

1 The table shows the weights of the women in a keep fit group.

Weight in kg (to the nearest kilogram)	Number of members
45–49	8
50–54	12
55–59	10
60–64	6
65–69	8

Write down the modal group.

2 The table shows the weekly pay of 30 workers.

Weekly pay (£)	Number of members
100–149.99	3
150–199.99	10
200–249.99	12
249–300.99	5

Write down the modal group.

3 The ages of the people at a bridge club are shown in this table.

Age	Number of people
20–29	5
30–39	12
40–49	8
50–59	15
60–69	26
70–79	10

Find the modal group.

4 The marks gained in a test are shown below.

55	60	62	74	53
59	73	81	91	48
43	62	90	85	45
63	67	75	49	84
61	67	44	68	77
83	49	84	76	63
53	73	88	64	74
79	56	63	44	59
82	59	65	57	87
83	72	70	51	48

Copy and complete the frequency table and write down the modal group.

Mark	Tally	Total
41–50		
51–60		
61–70		
71–80		
81–90		
91–100		

REVIEW EXERCISE 16H

1 The marks obtained by Jane in nine mathematics tests are

78 67 75 89 86
46 66 78 84

Calculate the mean value of these marks.

WJEC Summer 2005

2 Anika plays golf, her scores in her recent matches are:

75 74 69 83 79 82 65
87 75 73 71 74 81

(a) Find the range of the scores

(b) Find the mean value of these scores.

WJEC Summer 2004

3 Samantha is preparing for a holiday. She looks up the cost of the same holiday in a number of different brochures. The costs are shown below.

£320   £365   £299   £405
£354   £387   £368   £310

(a) Find the range of these costs.

(b) Find the mean value of these costs.

WJEC Summer 2003

4 The scores obtained by a player in eleven games of cricket are shown below.

15   0   35   2   56   2
75   5   21   2   17

(a) Write down the mode for these scores.

(b) Find the median value of the scores.

5 7 4 7 5 5 6 7 8 7 8 4 7 4 3 5

For the numbers shown above:

(a) find the median

(b) write down the mode

(c) find the range

(d) calculate the mean value

WJEC Summer 2002

6 23 35 33 26 53 28 45 35 18 44

For the numbers shown above:

(a) write down the range of the numbers

(b) calculate the mean value of the numbers.

WJEC Summer 2001

7 1 3 7 4 2 9 8 7 2 6 2 8 2 9 9

For the numbers shown above, write down

(a) the mode

(b) the median

WJEC Summer 2001

8 The numbers of computer games sold per day by a shop in the nine days before Christmas were as follows.

23   34   45   35   65
56   39   54   63

(a) What is the range of the number of computer games sold per day.

(b) Calculate the mean number of computer games sold per day.

WJEC Summer 2000

9   When the spinner shown below is
    spun, it is equally likely to land on any
    of the numbers shown.

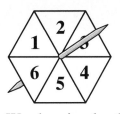

    Wendy spins the above spinner
    thirteen times. Her results are as
    follows:

    1   3   3   5   4   2   6
    6   4   5   5   3   3
    For these values, write down
    (i) the mode
    (ii) the median

WJEC Summer 2000

EXERCISE 17.1H

1 Find the first four terms of each of these sequences.
   (a) First term 3, term-to-term rule Add 5
   (b) First term −5, term-to-term rule Add 2
   (c) First term 23, term-to-term rule Subtract 5
   (d) First term 60, term-to-term rule Subtract 25
   (e) First term −17, term-to-term rule Add 6
   (f) First term −1, term-to-term rule Add $\frac{1}{2}$

2 Write down the next two terms in each of these sequences and give the term-to-term rule.
   (a) 1, 8, 15, 22, 29, 36, ...
   (b) 7, 13, 19, 25, 31, 37, ...
   (c) 9, 17, 25, 33, 41, 49, ...
   (d) 4, 13, 22, 31, 40, 49, ...
   (e) 16, 21, 26, 31, 36, 41, ...
   (f) 23, 30, 37, 44, 51, 58, ...

3 Write down the next two terms in each of these sequences and give the term-to-term rule.
   (a) 41, 35, 29, 23, 17, 11, ...
   (b) 39, 32, 25, 18, 11, 4, ...
   (c) 26, 21, 16, 11, 6, 1, ...
   (d) 29, 23, 17, 11, 5, −1, ...
   (e) 43, 35, 27, 19, 11, 3, ...
   (f) 68, 55, 42, 29, 16, 3, ...

4 Find the missing numbers in each of these sequences and give the term-to-term rule.
   (a) 1, 7, …, 19, 25, …
   (b) 11, …, 35, 47, …, 71
   (c) 85, …, 67, …, 49, 40
   (d) …, 87, 83, 79, …, 71
   (e) 7, 22, …, …, 67, 82
   (f) 34, …, 20, 13, 6, …

EXERCISE 17.2H

1 Draw the next pattern in each of these sequences.
   For each sequence, count the dots in each pattern and find the term-to-term rule.
   (a)
   (b)
   (c)
   (d)

2 Draw the next pattern in each of these sequences.
For each sequence, count the lines in each pattern and find the term-to-term rule.

(a)

(b)

(c)

(d)

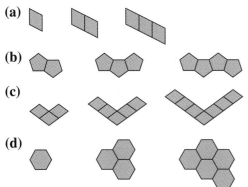

EXERCISE 17.3H

1 (a) Write down the next term of the sequence

3   8   13   18   23   ...

(b) Write down, in words, the rule for continuing the following sequence.

88   79   70   61   52   ...

WJEC Summer 2000

2 (a) Write down the next term in the sequence

4   8   12   16   20   ...

(b) Dots are used to make a sequence of patterns.
The first three patterns are shown below.

Draw the fourth pattern.

WJEC Summer 2001

3 (a) Write down the next two terms in the sequence

4   11   18   25   32   ...   ...

(b) Triangles, ▽, are used to make a sequence of patterns.
The first three patterns are shown below.

Draw the fourth pattern.

WJEC Summer 2002

4 Write down the next two numbers in the sequence

27,   25,   21,   15, ..., ...

WJEC Summer 2003

5 Write down the first three terms of the sequence whose nth term is $n^2 - 2$.

WJEC Summer 2004

6 (a) Write down the next term in the following sequence.

5   8   12   17   ....

(b) Dots are used to draw a sequence of patterns. The first three patterns are shown below.

Draw the next shape in the pattern.

WJEC Summer 2004

7 Dots are used to make a sequence of patterns.
The first three patterns are shown below.

```
                    •
        •         • •
  •    • •       • • •
```

Draw the **fifth** shape.

WJEC Summer 2005

8 96   48   24   12   6   …

   (i) Write down the next term in the above sequence.

   (ii) Describe, in words, the rule for finding the next term of the above sequence.

WJEC Summer 2005

18 → MENTAL METHODS 1

EXERCISE 18.1H

Work out these.

1	12 + 20	**2**	51 + 30
3	44 + 70	**4**	59 + 30
5	67 + 70	**6**	12 + 23
7	21 + 36	**8**	13 + 45
9	23 + 47	**10**	24 + 37
11	52 + 36	**12**	26 + 69
13	41 +23	**14**	49 + 36
15	36 +56	**16**	47 + 81
17	53 + 67	**18**	43 + 88
19	94 + 58	**20**	49 + 63

EXERCISE 18.2H

Work out these.

1	50 − 20	**2**	80 − 20
3	46 − 30	**4**	59 − 40
5	121 − 60	**6**	39 − 15
7	33 − 8	**8**	88 − 27
9	38 − 25	**10**	74 − 53
11	83 − 19	**12**	72 − 39
13	44 − 28	**14**	86 − 18
15	67 − 29	**16**	100−41
17	90 − 33	**18**	94 − 27
19	100 − 35	**20**	46 − 19

EXERCISE 18.3H

How many do you need to add to
1   25 to make 100?   **2** 63 to make 100?
3   27 to make 80?   **4** 42 to make 65?
5   51 to make 93?   **6** 37 to make 69?
7   29 to make 78?   **8** 46 to make 92?
9   25 to make 81?   **10** 56 to make 83?

EXERCISE 18.4H

Work out these.

1	2 × 3	**2**	3 × 2	**3**	7 × 6
4	6 × 7	**5**	7 × 2	**6**	4 × 8
7	9 × 9	**8**	5 × 4	**9**	3 × 3
10	8 × 3	**11**	9 × 6	**12**	5 × 8
13	3 × 7	**14**	6 × 4	**15**	9 × 10
16	8 × 8	**17**	10 × 3	**18**	5 × 7

EXERCISE 18.5H

Work out these.

1	8 ÷ 2	**2**	$\frac{24}{3}$	**3**	20 ÷ 5
4	24 ÷ 4	**5**	35 ÷ 5	**6**	$7\overline{)14}$
7	18 ÷ 6	**8**	42 ÷ 7	**9**	21 ÷ 3
10	$\frac{72}{9}$	**11**	30 ÷ 6	**12**	$\frac{72}{8}$
13	48 ÷ 6	**14**	$7\overline{)28}$	**15**	54 ÷ 9
16	$2\overline{)12}$	**17**	27 ÷ 3	**18**	60 ÷ 6

EXERCISE 18.6H

Estimate the value of each of these.

1	$\sqrt{19}$	**2**	$\sqrt{41}$	**3**	$\sqrt{74}$
4	$\sqrt{10}$	**5**	$\sqrt{26}$	**6**	$\sqrt{83}$
7	$\sqrt{97}$	**8**	$\sqrt{51}$	**9**	$\sqrt{42}$

EXERCISE 18.7H

1 Round these numbers to the nearest whole number.
   (a) 24.3   (b) 561.5
   (c) 81.2   (d) 417.6
   (e) 81.91   (f) 128.43
   (g) 39.8   (h) 627.203
   (i) 5.32   (j) 4.57

2 Round these numbers to 1 decimal place.
   (a) 2.37   (b) 8.24
   (c) 9.45   (d) 12.6666
   (e) 8.912   (f) 12.87
   (g) 9.624   (h) 7.465
   (i) 6.1919   (j) 1.95

3 Round these numbers to 2 decimal places.
   (a) 2.837   (b) 1.624
   (c) 5.465   (d) 5.1919
   (e) 1.2345   (f) 7.381
   (g) 6.247   (h) 7.318
   (i) 7.904   (j) 9.595

4 Round these numbers to 1 significant figure.
   (a) 64   (b) 372
   (c) 9414   (d) 98
   (e) 9012   (f) 54 217
   (g) 2594   (h) 45 092
   (i) 1631   (j) 1871.32

5 A man won £3 871 242 in a lottery. How much was this correct to 1 significant figure?

EXERCISE 18.8H

Work out these.
1 $3 - 5$ 2 $5 - 2$
3 $7 - 5$ 4 $4 - 6$
5 $3 - 4$ 6 $4 + 3$
7 $-1 + 3$ 8 $-3 + 3$
9 $-4 + 1$ 10 $2 + 1$
11 $-2 - 3$ 12 $-7 + 6$
13 $-3 + 2$ 14 $-6 - 5$
15 $-3 - 2$ 16 $-3 + (-4)$
17 $4 + (-5)$ 18 $-4 - (-3)$
19 $-2 + (-1)$ 20 $-6 - (-4)$

EXERCISE 18.9H

Work out these.
1 $5 - 3 + 2 - 1$
2 $3 + 3 - 5 - 2$
3 $2 + 3 + 2 - 5$
4 $6 - 4 - 3 + 2$
5 $-5 + 3 - 2 + 2$
6 $5 - 2 - 3 - 1$
7 $8 - 3 + 2 - 6$
8 $4 - 3 - 5 - 1 + 4 + 1$
9 $6 + 2 + 1 - 4 - 3 + 3$
10 $-2 - 4 + 2 + 8 - 6 - 2$

REVIEW EXERCISE 18H

1 (a) Write down, in figures, the number seven thousand and ninety seven.
  (b) 47   42   63   29   53   27   46
      (i) Write down two of the above numbers which add up to 100.
      (ii) Write down two of the above numbers whose difference is 13.
  (c) Write 57 259
      (i) to the nearest 100
      (ii) to the nearest 1000

WJEC Summer 2000

2 (a) Write down, in figures, the number seven thousand and eighty five.
   (b) (i) Work out $236 + 67$
       (ii) Work out $853 - 348$
   (c) Estimate the value of 48×7.9
       WJEC Summer 2001

3 (a) Write down, in figures, the number eight thousand and thirty seven.
   (b) Find the sum of 273 and 358.
   (c) Find the difference between 458 and 439.
   (d) Write 5838
       (i) correct to the nearest 10,
       (ii) correct to the nearest 100.
   (e) Andrew uses each of the digits 2 5 6 9 to make up four-digit numbers.
       (i) What is the smallest **even** number that he can make?
       (ii) What is the largest **odd** number that he can make?
   (f) Showing your working find an estimate for 49.8×4.9.
       WJEC Summer 2002

4 (a) Write down, in figures, the number six thousand seven hundred and eight.
   (b) 70   29   47   36   24
       52   28   63   53
       From the list of numbers shown above, write down
       (i) Two numbers which add up to 100
       (ii) Two numbers which have a difference of 46
   (c) Using all of the digits 2 9 5 7 write down the largest four digit number.
   (d) What is the value of the 6 in the number 7863?

   (e) The number 537 is multiplied by 10. What is the value of the 3 in the answer?
   (f) The number 6780 is divided by 10. What is the value of the 8 in the answer?
       WJEC Summer 2003

5 (a) (i) A television cost one thousand and eighty pounds. Write down the cost of the television in figures.
       (ii) Arnold wins £12,000,000. Write this amount in words.
   (b) (i) What must be added to 67 to make 94?
       (ii) What must be taken away from 105 to leave 48?
   (c) 912   921   936   953
       901   987   934   978
       From the above list of numbers write down
       (i) the smallest number,
       (ii) the largest number.
   (d) Showing all your working, estimate the value of 198.7×10.2
       WJEC Summer 2004

6 (a) Mark buys a suit and an overcoat. The suit cost £135 and the overcoat cost £129.
       Find the total cost of the suit and the overcoat.
   (b) Sian buys a dress.
       The cost of the dress is £124.
       Sian uses seven twenty pound notes to pay for the dress.
       How much change should Sian be given?

(c) Mary wants to buy a new camera.
She goes to five shops and notes
the price of the same camera in
each shop.
The prices are:
£278   £281   £298   £273   £297
Write down
   (i) the lowest price of the camera
   (ii) the highest price of the camera
                          WJEC Summer 2005

7   Only using numbers taken from the
    following list,
       2   5   9   12   17
      18   32   38   42

   (a) Write down the two numbers
       which add up to 49.
   (b) Write down the two numbers
       whose difference is 29.
   (c) Write down the two numbers
       which when multiplied together
       give an answer of 84.
   (d) When 63 is divided by a number
       the answer is 7.
       Write down the number.
                          WJEC Summer 2005

8   Use the fact that $86 \times 73 = 6278$ to
    write down the answers to the
    following
   (a) $8.6 \times 73 =$
   (b) $860 \times 7.3 =$
   (c) $6278 \div 73 =$
                          WJEC Summer 2005

EXERCISE 19.1H

1 For each of these shapes:
  • copy the shape on to squared paper.
  • draw an enlargement of the shape using the scale factor given.

(a) Scale factor 2

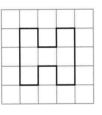

(b) Scale factor 3

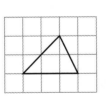

(c) Scale factor 3

(d) Scale factor 2

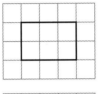

(e) Scale factor 2

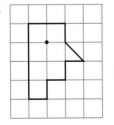

2 Work out the scale factor of enlargement of each of these pairs of shapes.

(a)

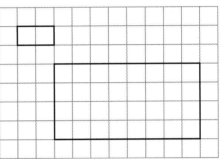

(b)

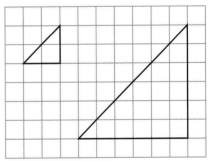

(c)

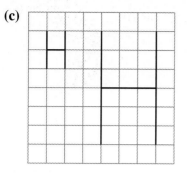

(d)

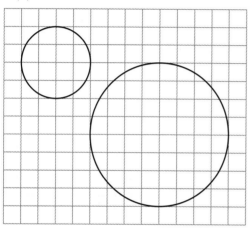

3 For each of these pairs of shapes, is the larger shape an enlargement of the smaller shape?
Give a reason for your answer.

(a)

(b)

EXERCISE 19.2H

1 Copy each of the shapes on to squared
paper.
Enlarge each of them by the scale
factor given.
Use the dot as the centre of the
enlargement.
(a) Scale factor 3

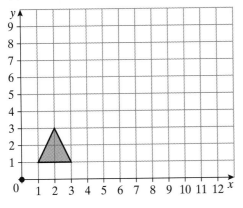

(b) Scale factor 2

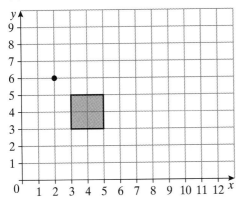

(c) Scale factor 4

(d) Scale factor 3

(e) Scale factor 2

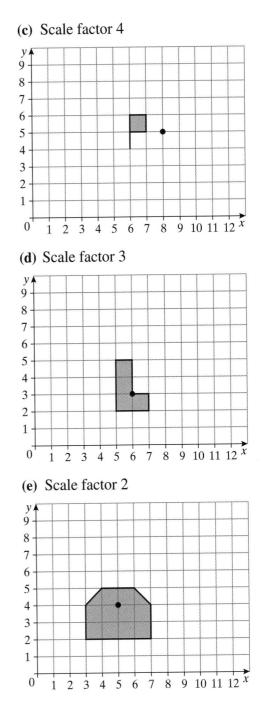

2 Copy each of these diagrams on to
   squared paper.
   For each of these diagrams find
   (i) the scale factor of the enlargement.
   (ii) the coordinates of the centre of the
        enlargement.

(a)

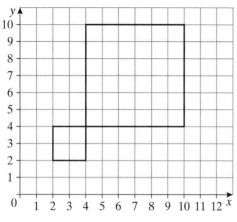

(b)

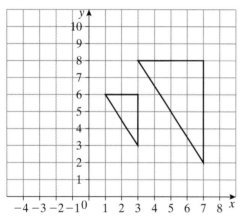

(c)

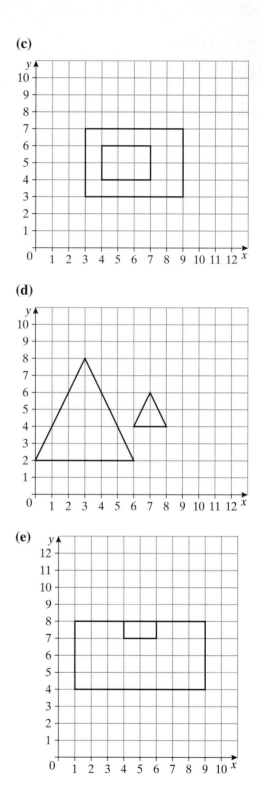

(d)

(e)

20 → PROBABILITY 1

EXERCISE 20.1H

1 Choose the best probability word from those below to complete these sentences.

> Impossible   Unlikely   Evens
> Likely   Certain.

(a) It is .... that the next baby born in Devon will be a boy.

(b) It is .... that there will be snow at the South Pole next week.

(c) It is .... that you will get more than 1 when you throw an ordinary dice.

2 Copy this scale.

Impossible Unlikely   Evens   Likely   Certain

Put arrows to show the chance of each of the following events happening.

(a) Water will come out when you turn on the tap.

(b) It will be dark tonight.

(c) Christmas Day falls on a Sunday.

3 There are 20 marbles in a tin. There are 3 red marbles, 10 green marbles and 7 yellow marbles.
A marble is taken out without looking. Choose the correct probability word to complete these sentences.

(a) It is .... that the marble is red.

(b) It is .... that the marble is blue.

(c) It is .... that the marble is green.

EXERCISE 20.2H

1 These six cards are laid face down and mixed up. Then a card is picked.

Copy this probability scale.

Put arrows to show the probability of each of the following events happening.
The number on the card is

(a) 3.

(b) less than 6.

(c) an even number.

2 There are ten pens in a bag.
Two are red, seven are blue and one is black.
A pen is taken out without looking.

(a) Use a probability word to complete this sentence.

> It is .... that the pen is blue.

(b) Copy this probability scale.

0   0.2   0.4   0.6   0.8   1.0

Mark the probability of each of these statements on your scale. Use arrows and label them R, G and B.
R: the pen is red.
G: the pen is green.
B: the pen is blue.

3 This spinner is fair.

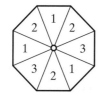

Calculate the probability of it landing on
(a) 3.
(b) an even number.
(c) 4.

4 An ordinary dice is thrown. Find the probability of getting
(a) an odd number.
(b) a 4.

5 Mary has 10 pens. Three are black, five are blue and two are green. She takes a pen without looking. Giving your answer as a decimal, calculate the probability that the pen is
(a) blue.
(b) green.
(c) not black.

6 There are 12 counters in a bag. The probability scale shows the probabilities of choosing the different colours of counter.

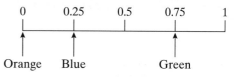

How many counters of each colour are there in the bag?

7 Gill has eight sweets in a bag. Three of them are yellow and the other five are pink.
(a) She takes a sweet without looking. Which colour is she more likely to get?
(b) Gill takes some sweets out of the bag.
There is now an even chance of getting a pink sweet.
How many sweets of each colour has she taken?

8 Jack has ten pairs of socks. Three of these pairs are black.
He takes a pair of socks out of his drawer without looking.
Draw a probability scale numbered 0 to 1, as in question **2**.
Mark the probability of each of these statements on your scale.
(a) The socks are black.
(b) The socks are not black.

9 The probability is $\frac{3}{8}$ that when three coins are tossed, you get two heads and a tail.
What is the probability that you don't get two heads and a tail?

10 There are five yellow counters, two red counters and eight black counters in a bag.
A counter is taken without looking. What is the probability that it is
(a) yellow?
(b) red?
(c) not red?

EXERCISE 20.3H

1 Irina surveyed the favourite crisp flavour of people in her class.
Here are her results.

Flavour	Frequency
Plain	7
Chicken	5
Cheese and Onion	12
BBQ	2
Other	4

Calculate an estimate of the probability that the next person she asks will like cheese and onion best.

2 Tim tests a spinner he has made to check it is fair
He spins it 500 times.
Here are his results.

Number on spinner	Frequency
1	58
2	73
3	103
4	124
5	142

(a) For this spinner, calculate the experimental probability of obtaining
   (i) a 5.   (ii) a 1.
(b) For a fair spinner, calculate, as a decimal, the probability of scoring
   (i) a 5.   (ii) a 1.
(c) Do your answers suggest that the spinner is fair? Give your reasons.

REVIEW EXERCISE 20H

1 A red bag contains five balls numbered as shown.

A blue bag contains six balls numbered as shown

① ③ ④ ⑥ ⑦ ⑧

In a game a player chooses a ball from the red bag and then a ball from the blue bag. The numbers on the two balls are added together to obtain a total score.

(a) Copy and complete the following table to show all the possible total scores.

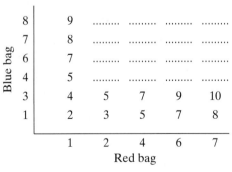

(b) (i) What is the probability that a player gets a total score of 14?

(ii) What is the probability that a player does not get a total score of 14?

A player wins a prize by getting a total score of 5 or less.

(c) (i) Tim plays the game once. What is the probability that he wins a prize?

(ii) 150 people each play the game once.
Approximately how many would you expect to win a prize?

(iii) It costs 20p to play the game once. The prize for scoring 5 or less is 40p. If the 150 people each play the game once, approximately how much profit do you expect the game to make?

WJEC Summer 2000

2   A square shaped spinner has the numbers 0, 1, 2 and 3 written on it. Another spinner, in the shape of a regular pentagon, has the numbers 1, 2, 3, 4 and 5 written on it.

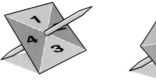

In a game, a player spins both spinners and multiplies the two numbers showing on the spinners to get the score for the game.

For example, if the number on the square spinner is 3 and the number on the pentagon spinner is 2, the player works out $3 \times 2 = 6$ and the player scores 6.

(a) Complete the following table to show all the possible scores.

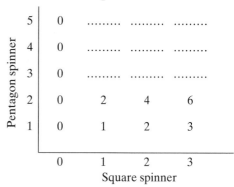

Pentagon spinner	Square spinner 0	1	2	3
5	0	....	....	....
4	0	....	....	....
3	0	....	....	....
2	0	2	4	6
1	0	1	2	3

(b) (i) What is the probability that a player scores 0?

(ii) What is the probability that a player does not score 0?

A player wins a prize by getting a score of 2 or less.

(c) Barbara plays the game once. What is the probability that she wins a prize?

(d) (i) 600 people each play the game once.
Approximately how many would you expect to win a prize?

(ii) It costs 30p to play the game once. The prize for getting a score of 2 or less is 50p. If the 600 people each play the game once, approximately how much profit do you expect the game to make?

WJEC Summer 2001

3 Jane throws a fair dice. On the scale given below mark the points A, B and C where,

A is the probability that the score will be a 5

B is the probability that the score will be an even number

C is the probability that the score will be an 8

```
0 ————————————+———————————— 1
```

WJEC Summer 2001

4 In a game, a player throws two fair dice, one coloured red the other blue.

The score for the throw is the smaller of the numbers showing. For example, if the red dice shows 5 and the blue dice shows 2, the score for the throw is 2; if the red dice shows 3 and the blue dice shows 3, the score for the throw is 3.

(a) Complete the following table to show all the possible scores.

Red dice						
6	1	2	3	....	....	....
5	1	2	3	....	....	....
4	1	2	3	....	....	....
3	1	2	3	3	....	....
2	1	2	2	2	....	....
1	1	1	1	1	1	1
	1	2	3	4	5	6

Blue dice

(b) (i) What is the probability that a player scores 1?

(ii) What is the probability that a player scores more than 1?

A player wins a prize by getting a score of 2 or less.

(c) William plays the game once. What is the probability that he wins a prize?

(d) (i) 360 people each play the game once. Approximately how many would you expect to win a prize?

(ii) It costs £1 to play the game once. The prize for winning is £1.50. If the 360 people each play the game once, approximately how much profit do you expect the game to make?

WJEC Summer 2002

5 (a) Using the probability scale below mark the points, **A**, **B** and **C** where,

A is the probability that there will be snow at the north pole in January,

B is the probability that it will rain in the Sahara desert tomorrow,

C is the probability that the score will be a 2 when an ordinary dice is rolled.

```
0 ——————————0.5——————————— 1
```

(b) One-hundred-and-fifty-five children in a school each buy one raffle ticket.

Wyn buys one of the raffle tickets.

(i) What is the probability of Wyn winning the raffle?

(ii) Wyn believes that the probability of the raffle being won by a boy is $\frac{1}{2}$. Is Wyn correct? Give a reason for your answer.

WJEC Summer 2002

6 ♠ ♦ ♠ ♣ ♣ ♠

Tim has six cards with shapes on them, as shown above.

(a) Tim chooses one card at random. Draw a circle around the shape he is most likely to choose?

♠   ♣   ♦

(b) On the probability scale shown below, mark the points A, B and C where,

A is the probability of Tim choosing a ♠

B is the probability of Tim choosing a ♦

C is the probability of Tim choosing a ♥

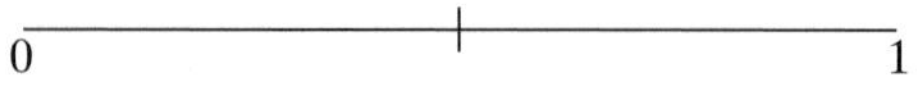

0 |————————+————————| 1

EXERCISE 21.1H

1 What are the readings on this scale?

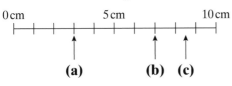

(a)   **(b)** **(c)**

2 What is the reading on this scale?

3 What are the readings on this thermometer?

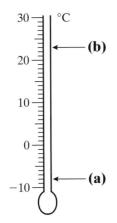

4 How long is this nail?

5 Measure the distance between A and B.

A ⊢———————————————⊣ B

EXERCISE 21.2H

1 Change these lengths to centimetres.
   (a) 2 m   **(b)** 3.5 m
   (c) 20 mm   **(d)** 15 m
   (e) 45 mm

2 Change these lengths to millimetres.
   (a) 2 cm   **(b)** 5.5 cm
   (c) 10 cm   **(d)** 2 m
   (e) 3.5 m

3 Change these weights to grams.
   (a) 2 kg   **(b)** 5 kg
   (c) 6.35 kg   **(d)** 0.8 kg
   (e) 0.525 kg

4 Put these volumes in order, smallest first.

   1.2 litres   500 ml   2 litres
   2500 ml   800 cl

5   Which metric units would you use to measure these lengths?
    (a) The span of your hand
    (b) The length of a corridor
    (c) The width of a window
    (d) The distance you can walk in a day
    (e) The distance from London to Edinburgh

6   Put these weights in order, smallest first.

    1.2 kg   1500 g   160 g
    2000 g   0.8 kg

7   Change these volumes to millilitres.
    (a) 2 litres   (b) 3.5 litres
    (c) 2 cl   (d) 15 cl
    (e) 0.345 litres

8   Graham has three pieces of string. The lengths are 45 cm, 85 mm and 1.2 m.
    (a) Write them down in order, shortest first.
    (b) What is the total length of string
        (i)   in millimetres?
        (ii) in centimetres?
        (iii) in metres?

EXERCISE 21.3H

Here are some approximate conversions between imperial and metric units.

Length	Weight
8 km ≈ 5 miles	1 kg ≈ 2.2 pounds (lb)
1 m ≈ 40 inches	
1 inch ≈ 2.5 cm	**Capacity**
1 foot (ft) ≈ 30 cm	4 litres ≈ 7 pints (pt)

1   Change these measures from imperial units to their approximate metric units.
    (a) 6 feet   (b) 35 miles   (c) 14 lb
    (d) 45 lb   (e) 14 pints

2   Change these measures from metric units to their approximate imperial units.
    (a) 24 km   (b) 5 m   (c) 5 kg
    (d) 12 litres (e) 20 cm

3   Stephen needs 1 pound of meat for a recipe.
    How many grams should he buy?

4   A radiator is 20 inches wide.
    What is this in centimetres?

5   Cola is sold in 2 litre bottles.
    How much is this in pints?

6   Pauline is 5 foot 4 inches tall.
    What is this in centimetres?

7   (a) A baby weighs 3.8 kilograms.
        Calculate the approximate weight of the baby in pounds.
    (b) 1.75 pints is approximately equal to 1 litre.
        A container holds 56 pints of paraffin.
        Calculate the approximate volume of the paraffin in litres.
                        WJEC Summer 2000

8   Mark drives towards Nottingham and sees this sign.

| Grantham | 23 miles |
| Nottingham | 53 miles |

    Roughly how far is it, in kilometres, from Grantham to Nottingham?

EXERCISE 21.4H

1 Estimate the following.
  (a) The height your desk
  (b) The width of a door
  (c) The mass of a glass of water
  (d) The capacity of a kitchen sink
  (e) The distance you can walk in an hour

2 Estimate the height of this lamp post.

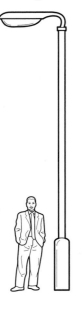

3 The building in this picture is about 16 m high.
  Estimate the height of the lorry.

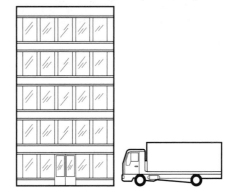

4 Alan wants to estimate the length of his bedroom. His foot is 20 cm long. He works out that the room is 16 foot lengths long.
  What is his estimate of the length of his bedroom?

5 Adrian is asked to estimate the distance he walks to school.
  His answer is 1.135 km.
  (a) Why is this not a sensible estimate?
  (b) What would be a better estimate?

EXERCISE 22.1H

1 This graph converts feet to centimetres.
   Use the graph to find how many
   (a) centimetres is 6 feet.
   (b) feet is 250 cm.
   (c) centimetres is 1 foot.

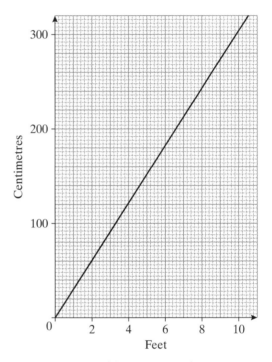

2 Area of land used to be measured in acres. It is now measured in hectares (ha).
   One hectare is 10 000 m². The graph converts between acres and hectares.

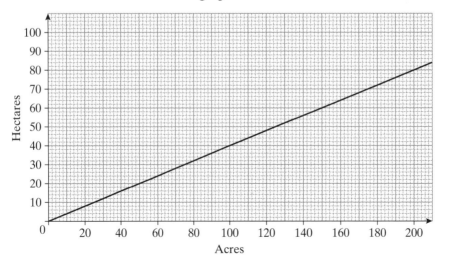

Use the graph to find how many
(a) hectares is 100 acres. (b) acres is 45 ha (c) acres is 1 ha.

EXERCISE 22.2H

1 Mr Smith is paying back a loan over a number of years.
The graph shows how much he owes.

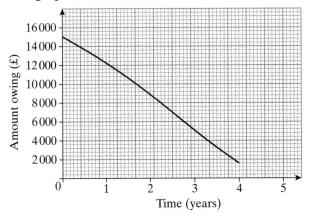

(a) How much did Mr Smith borrow?
(b) How much did he owe after
    (i) 2 years? (ii) 4 years?
(c) If he keeps on paying the money back at the same rate, when will he owe nothing?
    Give your answer to the nearest month.

2 The graph shows a patient's temperature over a day.

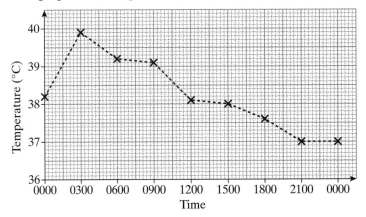

(a) When was the patient's temperature highest?
(b) What was the temperature at 1500?
(c) How often do you think the temperature was taken?
(d) Why are the points joined with a dotted line?

3 The graph shows the temperature of the water in a kettle, as it is heated.
   (a) After how many seconds is the water over 80 °C?
   (b) What is the temperature after 50 seconds?
   (c) How much does the temperature increase between 20 seconds and 40 seconds?

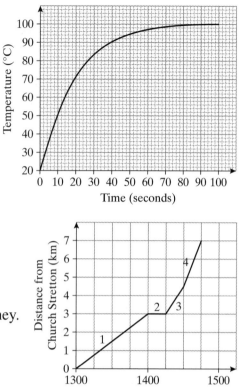

EXERCISE 22.3H

1 Jack is walking from Church Stretton to Ratlinghope across the Long Mynd, a hill in Shropshire.
   His journey is shown on the graph.
   Describe each of the four stages of Jack's journey.

2 Jasbir and Robert have a race over 2 km. Robert runs, then walks and then runs again. Jasbir jogs at a steady speed.
   The graph shows what happened.
   (a) Whose graph is the broken line?
   (b) Who won the race?
   (c) Describe what happened when the lines cross.

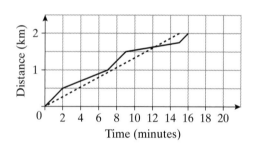

3 The graph shows two trains travelling between Leicester and St Pancras.
   The distance between the stations is 100 miles.
   (a) When did train B leave Leicester?
   (b) Which train stopped? How far from St Pancras was it when it stopped?
   (c) Find the time and distance from St Pancras when the trains passed each other.

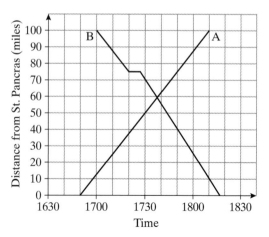

4 The graph shows Sue's journey by car from her home in Swansea to a services area and then on to Cheltenham, which is 120 miles from Swansea.

(a) How far did Sue travel in the first hour?

(b) For how many minutes did Sue stay at the service area?

(c) Without calculating any speeds, explain how you can tell whether Sue was, on average, travelling faster before or after her stop.

(d) Louise sets out from Cheltenham at 10.30 a.m. and travels at an average speed of 40 m.p.h. to the service area. Draw her journey on the graph paper.

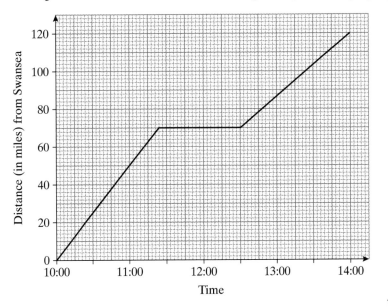

WJEC Summer 2004

EXERCISE 23.1H

1 The pictogram shows the number of customers buying pizzas in a takeaway restaurant one week.
   (a) How many pizzas were sold on Tuesday?
   (b) How many pizzas were sold on the busiest day?
   (c) How many pizzas were sold in the week?
   (d) What is the mean number of pizzas sold each day?

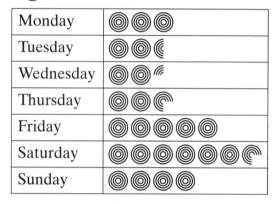

represents 20 sales

Monday	◎◎◎
Tuesday	◎◎◔
Wednesday	◎◎
Thursday	◎◎◔
Friday	◎◎◎◎◎
Saturday	◎◎◎◎◎◎◔
Sunday	◎◎◎◎

2 The pictogram shows the number of people attending a skating rink one week.
   (a) Which days were least popular for skating?
   (b) How many skaters were there on Wednesday?
   (c) How many skaters were there in the week?
   (d) What is the mean number of skaters each day?

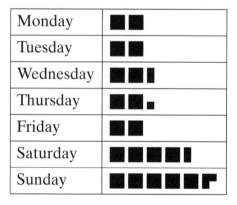

■ represents 40 skaters

Monday	■■
Tuesday	■■
Wednesday	■■▮
Thursday	■■▪
Friday	■■
Saturday	■■■■▮
Sunday	■■■■■◤

3 The pictogram shows the number of parasols sold by a garden centre one year.
   (a) How many parasols did the garden centre sell in summer?
   (b) How many parasols did the garden centre sell in the year?
   (c) What is the mean number of parasols sold each *month*?

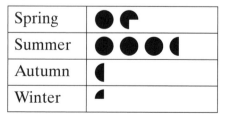

● represents 1000 parasols

Spring	●◗
Summer	●●◖
Autumn	◖
Winter	◤

4 The pictogram shows the number of cucumbers sold by a greengrocer one week.

(a) Which day of the week was the greengrocer closed?

(b) How many cucumbers were sold on Thursday?

(c) How many cucumbers were sold in the week?

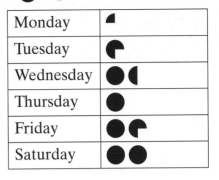

● represents 20 cucumbers

Monday	◤
Tuesday	◖
Wednesday	●◖
Thursday	●
Friday	●◖
Saturday	●●

5 The pictogram shows the number of drivers caught speeding by a speed camera one week.

(a) How many people were caught speeding on Wednesday?

(b) How many people were caught speeding on Monday?

(c) How many people were caught speeding in the week?

(d) What is the mean number of people caught speeding each day?

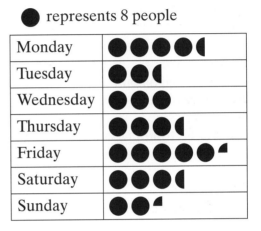

● represents 8 people

Monday	●●●●◖
Tuesday	●●◖
Wednesday	●●●
Thursday	●●◖
Friday	●●●●●◤
Saturday	●●●◖
Sunday	●●◤

6 The pictogram shows the number of televisions sold by a shop one year.

(a) How many televisions were sold in April?

(b) How many televisions are sold in the month with the least sales,

(c) How many televisions were sold in the year?

(d) What is the mean number of televisions sold each month?

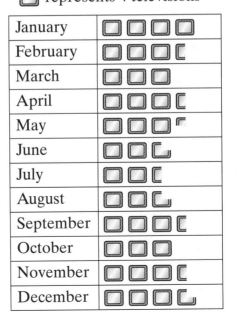

▢ represents 4 televisions

January	▢▢▢▢
February	▢▢▢◖
March	▢▢▢
April	▢▢▢◖
May	▢▢▢◜
June	▢▢◖
July	▢▢◖
August	▢▢◖
September	▢▢▢◖
October	▢▢▢
November	▢▢▢◖
December	▢▢▢◖

EXERCISE 23.2H

1 The pie chart shows the favourite flavour of crisp of 60 people questioned in a survey.

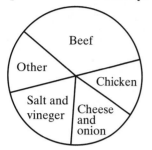

How many people preferred
(a) beef flavour?
(b) cheese and onion flavour?
(c) salt and vinegar flavour?

2 The pie chart shows the eye colour of 108 people questioned in a survey.

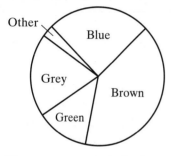

How many people had
(a) blue eyes?   **(b)** grey eyes?

3 The pie chart shows the distribution of the population of Britain.

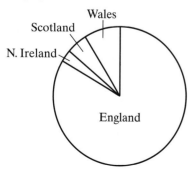

The total population of Britain is approximately 60 million people. What is the approximate population of
(a) England?   **(b)** Wales?
(c) Northern Ireland?
(d) Scotland?

4 The pie chart shows the approximate land area of Britain.

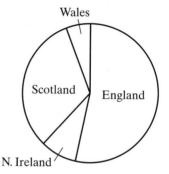

The total land area of Britain is approximately 94 320 square miles. What is the approximate land area of
(a) England?   **(b)** Wales?
(c) Northern Ireland?
(d) Scotland?

5 The pie chart shows the ingredients needed to make fruit crumble.

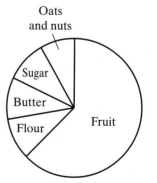

Betty makes a large fruit crumble weighing 2.4 kg.
What is the approximate weight, in grams, of each of the ingredients of Betty's crumble?

EXERCISE 23.3H

1 The temperature of some water was taken every 5 minutes as it was heated.

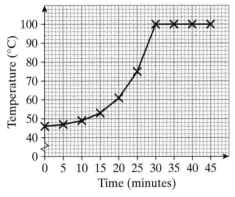

(a) What was the temperature of the water before it was heated?

(b) How many minutes did it take for the liquid to reach 75°C?

(c) How many degrees did the water temperature rise in the first 20 minutes?

(d) What happened after 30 minutes?

2 The line graph shows the number of spectators at the first six games played by Rangers and by Rovers in the season.

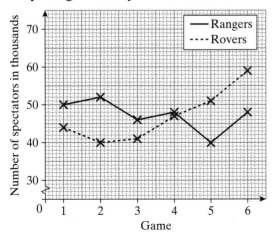

(a) How big was the crowd for at Rovers' second game?

(b) For which game did both teams have approximately the same size crowd?

(c) What was the range of the crowd size for the Rangers games?

(d) What was the mean crowd size for the Rovers games?

(e) Can you use this information to predict crowd sizes for game 7? Explain your answer.

3 The line graph below shows the total rainfall in Bananaland theme park one year.

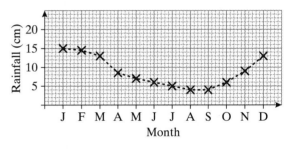

(a) How much rain fell in June?

(b) Which month had the most rain?

(c) What was the range of the monthly rainfall figures for the year?

(d) Calculate the mean monthly rainfall for the year.

4 The line graph shows the value of sales of a mushroom farm one year.

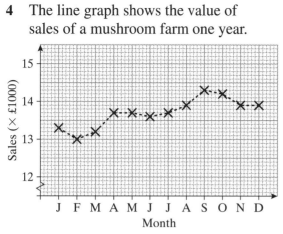

(a) Which month had the lowest sales?

(b) What was the range of sales values for the year?

(c) What was the total value of sales for the year?

(d) What was the mean value of sales each month?

5 The line graph shows the number of drivers paying to cross a toll bridge one week.

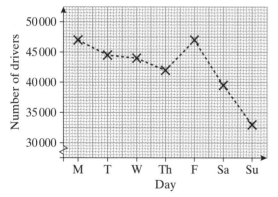

(a) How many drivers crossed the bridge during the week?

(b) What was the mean number of drivers crossing the bridge each day?

(c) What was the range of the number of drivers crossing the bridge each day?

(d) Why might there be fewer drivers crossing the bridge on Saturday and Sunday?

6 The line graph shows the mean monthly rainfall and mean maximum daily temperature in Brisbane one year.

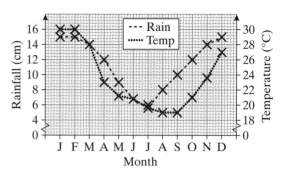

(a) Which month had the highest mean monthly rainfall?

(b) Which months had the highest maximum daily temperature?

(c) What was the maximum daily temperature in September?

(d) In which month was the mean monthly rainfall approximately 14 cm?

(e) What was the range of the maximum daily temperature for the year?

(f) What was the range of the mean monthly rainfall for the year?

EXERCISE 23.4H

1 The pie charts show the number of companies involved in the automotive trade in the UK in 1998 and 2001. The total number of companies in each year was the same.

Number of companies in 1998

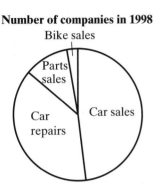

Number of companies in 2001

(a) In which year was the number of car repair companies larger?

(b) What is the main difference between the two years?

(c) What similarities are there between the two years?

2 The diagrams below show the performances of boys and girls in a general knowledge test.

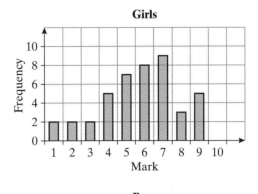

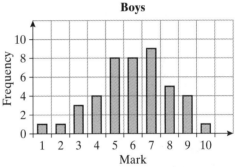

(a) How many girls scored 7 in the test?

(b) How many boys scored 4 in the test?

(c) What was the modal score for the girls?

(d) What was the modal score for the boys?

(e) What was the range of the scores for the girls?

(f) Did girls or boys do better in this test?
Give a reason for your answer.

3 One measure of the health of a country is its Infant Mortality Rate (IMR). The bar charts below show the IMR of five developed and five undeveloped countries.

Developed

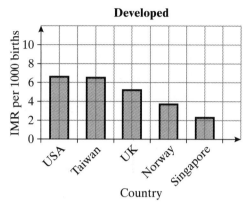

Undeveloped

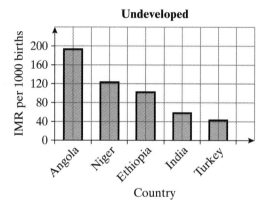

(a) Approximately, what is the IMR of the UK?

(b) Approximately, what is the IMR of Angola?

(c) Approximately, what is the mean IMR for the undeveloped countries?

(d) Approximately, what is the mean IMR for the developed countries?

(e) Approximately, what is the range of the IMRs for all 10 countries?

(f) Which group of countries is the healthiest?

(g) Draw a bar chart to show the data for all ten countries.

EXERCISE 24.1H

1   Find the perimeter of each of these shapes.

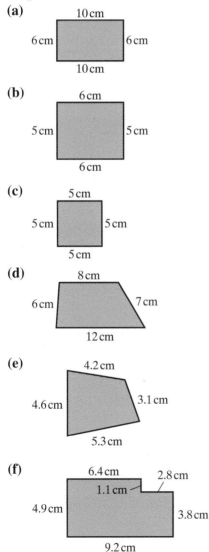

(a)

10 cm

6 cm   6 cm

10 cm

(b)

6 cm

5 cm   5 cm

6 cm

(c)

5 cm

5 cm   5 cm

5 cm

(d)

8 cm

6 cm   7 cm

12 cm

(e)

4.2 cm

4.6 cm   3.1 cm

5.3 cm

(f)

6.4 cm   2.8 cm

1.1 cm

4.9 cm   3.8 cm

9.2 cm

2   A square has sides of length 3.7 m. What is its perimeter?

3   A rectangle has sides of length 4.3 cm and 5.7 cm. What is its perimeter?

4   The front of a calculator is a rectangle with sides of length 13.5 cm and 6.5 cm. What is its perimeter?

5   The top of a filing cabinet is a rectangle with sides of length 60 cm and 45 cm. What is its perimeter?

EXERCISE 24.2H

1 Find the area of each of these shapes.
Give your answers in cm².

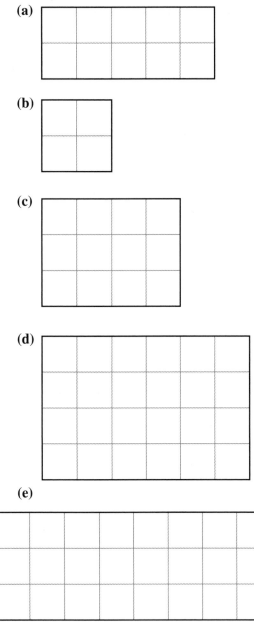

(a)

(b)

(c)

(d)

(e)

EXERCISE 24.3H

1 Find the area of each of these rectangles.
Take care to give the correct units in
the answer.

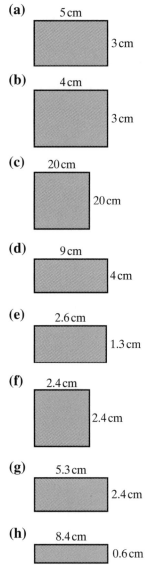

(a) 5 cm, 3 cm

(b) 4 cm, 3 cm

(c) 20 cm, 20 cm

(d) 9 cm, 4 cm

(e) 2.6 cm, 1.3 cm

(f) 2.4 cm, 2.4 cm

(g) 5.3 cm, 2.4 cm

(h) 8.4 cm, 0.6 cm

2 A rectangle measures 5.3 cm by
2.6 cm. Find its area.

3 A square has sides of length 3.6 cm.
Find its area.

4 A rectangle has sides of length 3.25 cm and 0.95 cm. Find its area.

5 A rectangular field measures 35 m by 16 m. Find its area.

6 A desk top is a rectangle measuring 1.2 m by 0.8 m. Find its area.

7 A drive is a rectangle measuring 6.5 m by 2.8 m.
   (a) Find the area of the drive.
   To pave the drive it costs £32.50 per square metre.
   (b) How much does it cost to pave the drive?

8 A rectangular window measures 1.5 m by 3.5 m.
   (a) Find the area of the window.
   Glass costs £25 per square metre.
   (b) How much does the glass for the window cost?

EXERCISE 24.4H

1 Find the volume of each of these cuboids.
   (a)
   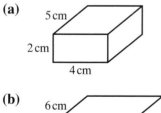
   5 cm
   2 cm
   4 cm

   (b)
   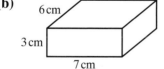
   6 cm
   3 cm
   7 cm

2 A cuboid has a height of 6 cm, a length of 4 cm and a width of 5 cm. Find its volume.

3 A cube has edges 4 cm long. Find its volume.

4 A cuboid has a square base with sides of length 4.5 cm and is 6 cm high. Find its volume.

5 Calculate the volume of a cube with edges 12 m long.

6 A camera box has a rectangular base measuring 16 cm by 14 cm and is 12 cm high.
Calculate its volume.

7 A fish tank has a rectangular base measuring 15 cm by 60 cm long and is 25 cm deep.
Find the volume of the tank.

8 A box of chocolates is a cuboid 12.5 cm long, 6 cm wide and 4.6 cm deep.
Calculate its volume.

9 A biscuit tin is a cuboid 12 cm long, 4.5 cm wide and 6.5 cm deep.
Work out its volume.

10 A piece of wood is a cuboid 3 m wide, 60 cm long and 2.5 cm thick.
Find the volume of the wood.
Hint: Be careful with the units.

EXERCISE 25.1H

1 Work out these squares.

(a) 4.3^2 (b) 7.2^2

(c) 56^2 (d) 419^2

(e) 0.74^2 (f) 0.82^2

(g) 0.09^2 (h) 3.2^2

(i) 4.71^2 (j) 63.8^2

2 Work out these square roots.

(a) $\sqrt{14.44}$ (b) $\sqrt{68.89}$

(c) $\sqrt{3.61}$ (d) $\sqrt{219.04}$

(e) $\sqrt{3844}$ (f) $\sqrt{0.1156}$

(g) $\sqrt{0.5776}$ (h) $\sqrt{0.0324}$

(i) $\sqrt{10.3041}$ (j) $\sqrt{0.007\,225}$

3 Work out these square roots. Give your answers correct to 2 decimal places.

(a) $\sqrt{45.3}$ (b) $\sqrt{74.8}$

(c) $\sqrt{44}$ (d) $\sqrt{827}$

(e) $\sqrt{5632}$ (f) $\sqrt{0.468}$

(g) $\sqrt{0.4}$ (h) $\sqrt{56\,846}$

(i) $\sqrt{0.408}$ (j) $\sqrt{0.063}$

4 The area of a square is 720 cm². Find the length of the side of the square. Give your answer correct to 1 decimal place.

EXERCISE 25.2H

Work these out on your calculator without writing down the answers to the middle stages.

If the answers are not exact, give them correct to 3 decimal places.

1 $(7.3 + 3.2) \div 4.8$

2 $(134 - 43) \div 35$

3 $(8.2 - 3.6) \times 5.4$

4 $\sqrt{(17.3 + 16.8)}$

5 $\sqrt{(68.7 - 2.3^2)}$

6 $(7.3 - 2.6)^2$

7 $7.3^2 - 2.6^2$

8 $16.8 \div (5.2 - 1.9)$

9 $5.8 \times (1.9 + 7.3)$

10 $\sqrt{(28.6 - 9.7)}$

11 $\sqrt{(26.2 \div 3.8)}$

12 $34.9 \div (2.8 + 5.3)$

EXERCISE 26.1H

1 Measure the length of each of these lines in centimetres.

(a) ────────────────────────────

(b) ──────────

(c) ─────────────────────

(d) ────────────

(e) ──────────────────────

2 Measure the length of each of these objects.

(a)

(b)

3 Measure the length of each side of these shapes.

(a)　　　　　　　　　　　　　　　　(b)

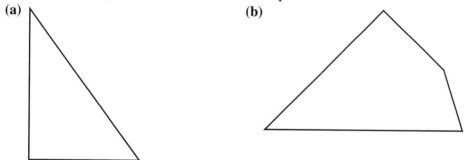

EXERCISE 26.2H

Copy and complete this table.
Estimate each angle first and then
measure it with a protractor.

Angle	Estimated size	Measured size
a		
b		
c		
d		
e		
f		
g		
h		
i		
j		

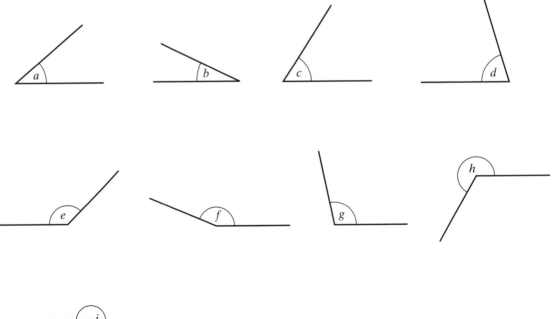

EXERCISE 26.3H

1 Draw accurately each of these angles.

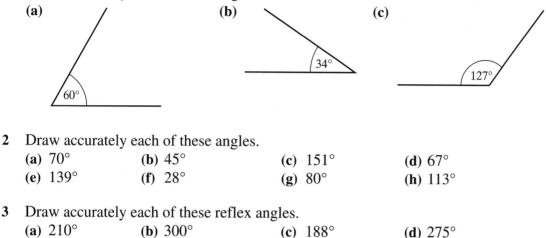

(a)

60°

(b)

34°

(c)

127°

2 Draw accurately each of these angles.

   (a) 70°   **(b)** 45°   **(c)** 151°   **(d)** 67°

   (e) 139°   **(f)** 28°   **(g)** 80°   **(h)** 113°

3 Draw accurately each of these reflex angles.

   (a) 210°   **(b)** 300°   **(c)** 188°   **(d)** 275°

EXERCISE 26.4H

1 Make an accurate full-size drawing of each of these triangles.
For each triangle, measure the unknown length and angles from your drawing.

(a)

B

4 cm

38°

A   5 cm   C

(b)

N

147°   5.5 cm

L   6.2 cm   M

(c) Triangle ABC where AB = 8 cm, angle BAC = 58° and AC = 5 cm.

2 Make an accurate full-size drawing of each of these triangles.
For each triangle, measure the unknown lengths and angle from your drawing.

(a)

R

60°   53°

P   6 cm   Q

(b) C

100°   27°

A   5.4 cm   B

(c) Triangle XYZ where YZ = 6.5 cm, angle XZY = 67° and angle ZYX = 43°.

EXERCISE 26.5H

1 Make an accurate full-size drawing of each of these triangles.
   For each triangle, measure all the angles from your drawing.

   (a)   R   5 cm   6 cm   P   7 cm   Q

   (b)   I   4 cm   8 cm   G   6 cm   H

   (c) Triangle ABC where AB = 6 cm, BC = 6.5 cm and AC = 7 cm.

2 Make an accurate full-size drawing of each of these triangles.
   For each triangle, measure the unknown length and angles from your drawing.

   (a) N   9 cm   L   7 cm   M

   (b) F   8 cm   100°   D   4.5 cm   E

   (c) Triangle XYZ where YZ = 6.8 cm, angle XZY = 42° and XY = 8 cm.

EXERCISE 26.6H

1 Measure each of these lines as accurately as possible.
   Using the scales given, work out the actual length that each line represents.

   (a) —————————————————   1 cm to 6 m

   (b) ————————————   1 cm to 20 km

   (c) ———————————————————   2 cm to 5 miles

   (d) ———————————   1 cm to 4 m

2 Draw accurately the line to represent these actual lengths.
   Use the scale given.
   (a) 6 m   Scale: 1 cm to 1 m
   (b) 8 km   Scale: 1 cm to 2 km
   (c) 15 miles   Scale: 3 cm to 5 miles
   (d) 450 m   Scale: 1 cm to 100 m

3 Here is a plan of a bedroom.
The scale of the drawing is
1 cm to 50 cm.

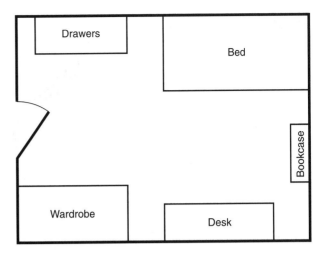

(a) Work out the actual length and width of the bedroom.

(b) Work out the actual length and width of each of the five items in the bedroom.

(c) The window in the bedroom measures 1 m by 1 m 75 cm.
What will the measurements of the window be on this scale drawing?

4 The map shows some towns and cities in Scotland.
The scale of the map is 1 cm to 10 km.

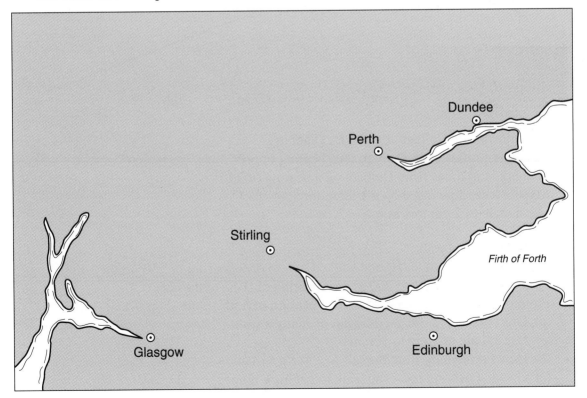

(a) What is the real-life distance, in kilometres, between these towns?
- **(i)** Glasgow and Stirling
- **(ii)** Edinburgh and Glasgow
- **(iii)** Edinburgh and Perth
- **(iv)** Glasgow and Dundee
- **(v)** Perth and Dundee
- **(vi)** Edinburgh and Stirling

(b) It is 660 km from Edinburgh to London.
How many centimetres will this be on the map?

EXERCISE 26.7H

1 The diagram shows a section of a plan of a school. From a point P in the playground the plan shows the position of the entrances to various schoolrooms.
Find the bearing of each of the entrances from the point P.
The scale of the map is 1 cm to 10 m.
Find the actual distance of each of the entrances from P.

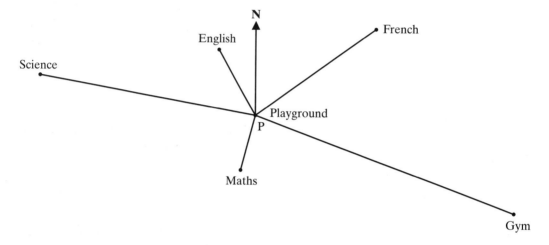

2 Three towns are Delham, Easton and Fanbury.
Easton is 3 miles from Delham on a bearing of 065°.
Fanbury is $2\frac{1}{2}$ miles from Delham on a bearing of 290°.
Make a scale drawing showing these three towns.
Use a scale of 2 cm to 1 mile.

3 Jenny went for a walk.
She started from home (H) and walked 500 m on a bearing of 150° to the zoo (Z).
From the zoo she walked 900 m on a bearing of 088° to the cafe (C).
From the café she walked 600 m on a bearing of 350° to the shop (S).
- **(a)** Draw an accurate scale drawing of Jenny's walk.
  Use a scale of 1 cm to 100 m.
- **(b)** How far is Jenny from home?
- **(c)** On what bearing must she walk to get back home?

EXERCISE 27.1H

1 How many hours and minutes are there between the following times?
   (a) 08:10 and 08:43
   (b) 15:00 and 19:00
   (c) 10:42 and 11:23
   (d) 13:48 and 15:22
   (e) 06:41 and 14:25
   (f) 22:10 on Thursday and 04:20 on Friday
   (g) $\frac{1}{4}$ to 4 and $\frac{1}{2}$ past 5
   (h) 20 minutes to 5 and 25 minutes past 8

2 Felicity arrived at the dentist at 12:40. She was at the dentist for 35 minutes. What time did she leave?

3 Gary went to town on a bus. He arrived at the bus stop at 09:55 and waited 8 minutes for the bus. The bus took 23 minutes to get to town. What time did Gary arrive in town?

4 (a) On weekdays a train leaves Sheffield for London at 13:27 and arrives in London at 15:50. How long does the journey take?
   (b) On Sundays a train leaves Sheffield for London at 12:51. It takes 3 hours and 13 minutes. At what time does the train arrive in London?

5 James has an appointment for 10:15. He arrives 16 minutes early. What time does he arrive?

6 At what time do these TV programmes finish?
   (a) Starts at 08:15 and lasts 30 minutes.
   (b) Starts at 12:20 and lasts 50 minutes.
   (c) Starts at 18:25 and lasts 40 minutes.
   (d) Starts at 20:33 and lasts 1 hour 35 minutes.
   (e) Starts at 22:35 on Tuesday and lasts 2 hours 20 minutes.

7 Rupinder goes for lunch at 12:35 and returns at 13:47. How long is she away?

8 Adam arrives 8 minutes early for a meeting due to start at 19:30. It starts 3 minutes late and finishes at 21:18.
   (a) What time did Adam arrive?
   (b) How long was the meeting?

9 Rachel runs the 10 000 m race in 38 minutes 14 seconds. Caroline runs it in 39 minutes and 2 seconds. How much longer did it take Caroline?

10 Emily runs in a marathon. She checks the exact time when she crosses the start line. It is 10:41:05. She completes the race in a time of 4 hours 13 minutes and 6 seconds. What time is it when she finishes?

EXERCISE 27.2H

1 Kieran drove 180 kilometres at an average speed of 40 km/h.
How long did he take?

2 Imogen walked at an average speed of 5 km/h for 3 hours 30 minutes.
How far did she walk?

3 Patrick drove from his home to London, a distance of 189 miles.
He took four and a half hours.
What was his average speed in miles per hour?

4 A car travels at an average speed of 84 km/h.
How far does it travel in 2.6 hours?

5 It took Peter 4 minutes 10 seconds to run 1000 metres.
What was his speed in metres per second.

6 Mandy went on a ride on her motor bike.
Her average speed was 96 km/h and she took 1 hour 40 minutes.
How far did she travel?

7 A plane takes 3 hours 30 minutes to fly 2611 km.
What was its average speed?

8 The distance from Lake Louise to Radium Springs is 240 miles.
Taj travelled at an average speed of 54 mph.
How long did it take him? Give your answer correct to the nearest minute.

9 A long cycle race covers 275 km. The winner took 14 hours 15 minutes.
What was his average speed? Give your answer correct to 1 decimal place.

10 (a) How long does it take to travel 100 miles at an average speed of 70 mph?
   (b) How much time do you save by travelling 10 mph faster?
   Give your answers correct to the nearest minute.

EXERCISE 27.3H

Use these approximate conversions for changing between imperial and metric units.

Length	Weight
8 km ≈ 5 miles	1 kg ≈ 2 pounds (lb)
1 m ≈ 40 inches	25 g ≈ 1 ounce (oz)
1 inch ≈ 2.5 cm	**Capacity**
1 foot (ft) ≈ 30 cm	4 litres ≈ 7 pints (pt)

1 Petra went on a 12 km hike.
How far was this in miles?

2 For a party Jasmine bought a 2 litre carton of full cream milk and four 2 litre bottles of semi-skimmed milk.
How many pints of milk is that altogether?

3 Simon's rucksack has a mass of 32 pounds.
How much is this in kilograms?

4 Fitz bought four lengths of wood, each 8 foot long.
What was the total length in metres and centimetres?

5 At the swimming baths there is a 5-metre diving board.
How high is this in feet and inches?

6 The distance from Sheffield to Cambridge is 185 miles via the M1, but only 148 miles if you cut across to the A1.
How much further, in kilometres, is the journey via the M1?

7 My car holds 10 gallons of petrol (1 gallon = 8 pints).
How many litres does it hold? Give your answer to the nearest litre.

8 On a flight passengers are each allowed to have luggage weighing up to 20 kg.
Use the more accurate conversion 1 kg = 2.2 lbs to find how much they are each allowed in pounds.

EXERCISE 27.4H

1 On holiday in France, Jason made this table of approximate conversions between kilometres per hour (km/h) and miles per hour (mph).

km/h	40	50	60	70	80	90	100	110	120
mph	25	30	40	45	50	55	60	65	75

A more accurate conversion is mph = km/h × 0.625.
(a) Jason saw a speed limit sign of 90 km/h.
What is the speed limit in mph
(i) according to his table?
(ii) using the more accurate conversion?

(b) Some of the conversions in the table are exactly the same as using the accurate conversion.
List them.

2 Here is part of a Sunday timetable for trains from Sheffield and Chesterfield to London St Pancras.

Sheffield	Chesterfield	London St Pancreas
0853	0906	1219
0951	1007	1318
1102	1113	1418
1251	1303	1604
1353	1407	1706
1443	1456	1755
1601	1616	1914
1830	1843	2141

(a) What time does the 1102 from Sheffield arrive at London St Pancras?
(b) What time does the train arriving in London at 1755 leave Chesterfield?
(c) Which of the trains in the list completes the journey from Sheffield to London in the shortest time?

3

Distances are in miles	Birmingham	Bristol	Edinburgh	Glasgow	Liverpool	London	York
Birmingham		87	286	287	90	110	127
Bristol	87		364	365	159	116	214
Edinburgh	286	364		44	210	373	184
Glasgow	287	365	44		212	392	207
Liverpool	90	159	210	212		197	97
London	110	116	373	392	197		196
York	127	214	184	207	97	196	

Use the above mileage chart to find:

(i) the distance from **Bristol** to **London**,

(ii) the **total length** of a journey from **Glasgow** to **Edinburgh** to **York**.

WJEC Summer 2004

4   The following timetable shows the times of buses from Cefn Park to Church Road.
The buses then return from Church Road to Cefn Park.

Cefn Park – Church Road – Cefn Park	Bus Timetable			
	Service numbers			
	2901	**2902**	**2903**	**2904**
Cefn Park	09 32	10 45	12 12	14 35
Kings Road		10 51	12 18	14 41
Saint Mary Street		10 58	12 25	14 48
Castle Hill		11 04	12 31	14 54
Church Road	09 52	11 11	12 38	15 01
Castle Hill	09 59	11 18		15 08
Saint Mary Street	10 07	11 26		15 16
Kings Road	10 15	11 34		15 24
Cefn Park	10 24	11 43	12 58	15 33

(a) What time does service 2902 leave Cefn Park?
   At what **times** does service 2904 arrive at Kings Road?

(b) How long does it take service 2903 to travel from Cefn Park to Church Road and
   return to Cefn Park?   **WJEC Summer 2005**

5 (a)

Distances are in miles	London	Cardiff	Fishguard	Liverpool	Shrewsbury
London		152	263	203	156
Cardiff	152		127	198	110
Fishguard	263	127		167	132
Liverpool	203	198	167		68
Shrewsbury	156	110	132	68	

Using the above mileage chart to find
    (i) the distance from **Liverpool** to **Cardiff**,
    (ii) the total length of the journey from **London** to **Cardiff** to **Fishguard**.
(b) The following table is a section of the rail timetable from Swansea to London.

Swansea	07 00	10 30	13 30
Neath	07 10	10 40	13 40
Port Talbot	07 19	10 49	13 49
Bridgend	07 30	11 00	14 00
Cardiff	07 55	11 25	14 25
Newport	08 15	11 45	14 45
Bristol Parkway	08 40	12 10	15 10
Reading	09 26	13 01	16 01
London	09 56	13 30	16 30

    (i) How long should it take the 07 00 train from **Swansea** to get to **Port Talbot**?
    (ii) How long should it take the 13 30 train from **Swansea** to travel from **Cardiff** to **Bristol Parkway**?

WJEC Summer 2005

28 → INTEGERS, POWERS AND ROOTS 2

EXERCISE 28.1H

Write each of these numbers as a product of its prime factors.

1	14	**2**	16
3	28	**4**	35
5	42	**6**	49
7	108	**8**	156
9	225	**10**	424

EXERCISE 28.2H

For each of these pairs of numbers
- write the numbers as products of their prime factors.
- state the highest common factor.
- state the lowest common multiple.

1	6 and 8	**2**	8 and 18
3	15 and 25	**4**	36 and 48
5	25 and 55	**6**	33 and 55
7	54 and 72	**8**	30 and 40
9	45 and 63	**10**	24 and 50

EXERCISE 28.3H

Work out these.

1	2×3	**2**	-5×8
3	-6×-2	**4**	-4×6
5	5×-7	**6**	-3×7
7	-4×-5	**8**	$28 \div -7$
9	$-25 \div 5$	**10**	$-20 \div 4$
11	$24 \div 6$	**12**	$-15 \div -3$
13	$-35 \div 7$	**14**	$64 \div -8$
15	$27 \div -9$	**16**	$3 \times 6 \div -9$

17 $-42 \div -7 \times -3$
18 $5 \times 6 \div -10$
19 $-9 \times 4 \div -6$
20 $-5 \times 6 \times -4 \div -8$

EXERCISE 28.4H

Do not use your calculator for questions **1** and **2**.

1 Write down the value of each of these.
   (a) 1^2 (b) 13^2 (c) $\sqrt{64}$
   (d) $\sqrt{196}$ (e) 3^3 (f) 5^3
   (g) $\sqrt[3]{8}$ (h) $\sqrt[3]{64}$

2 A cube has sides of length 4 cm. What is its volume?

You may use your calculator for questions **3** to **7**.

3 Find the square of each of these numbers.
   (a) 20   (b) 42   (c) 5.1
   (d) 60   (e) 0.9

4 Find the cube of each of these numbers.
   (a) 7   (b) 3.5   (c) 9.4
   (d) 20   (e) 100

5 Find the square root of each of these numbers.
   Where necessary, give your answer correct to 2 decimal places.
   (a) 900   (b) 75   (c) 284
   (d) 31 684   (e) 40 401

6 Find the cube root of each of these numbers.
Where necessary, give your answer correct to 2 decimal places.
(a) 729   (b) 144   (c) 9.261
(d) 4848   (e) 100 000

7 A square has an area of 80 cm².
What is the length of one side? Give your answer correct to 2 decimal places.

EXERCISE 28.5H

1 Write these in simpler form using indices.
(a) $2 \times 2 \times 2 \times 2 \times 2 \times 2$
(b) $7 \times 7 \times 7 \times 7$
(c) $2 \times 2 \times 3 \times 3 \times 3 \times 3 \times 5 \times 5 \times 5$

2 Work out these giving your answers in index form.
(a) $2^2 \times 2^4$ (b) $3^6 \times 3^2$
(c) $4^2 \times 4^3$ (d) $5^6 \times 5$

3 Work out these giving the answers in index form.
(a) $5^5 \div 5^2$ (b) $7^8 \div 7^2$
(c) $2^6 \div 2^4$ (d) $3^7 \div 3^3$

4 Work out these giving your answers in index form.
(a) $\dfrac{2^5 \times 2^4}{2^3}$ (b) $\dfrac{3^7}{3^5 \times 3^2}$
(c) $\dfrac{5^5 \times 5^4}{5^2 \times 5^3}$ (d) $\dfrac{7^5 \times 7^2}{7^2 \times 7^4}$

5 (a) Write each of these numbers as a product of its prime factors.
(i) 36   (ii) 49   (iii) 64
(iv) 100   (v) 324

(b) All the numbers in part (a) are square numbers. Write down what you notice about all the powers (indices) in (a).

6 Write 50 as a product of its prime factors. What is the least number that 50 needs to be multiplied by so that the result is a square number?

7 (a) Write each of these numbers as a product of its prime factors.
(i) 12   (ii) 27   (iii) 60
(iv) 75   (v) 112
(b) For each number in parts (i) to (v), find the least number it needs to be multiplied by so that the result is a square number?

EXERCISE 28.6H

Do not use your calculator for questions **1** to **3**.

1 Write down the reciprocal of each of these numbers.
(a) 4   (b) 9   (c) 65

2 Write down the numbers of which these are the reciprocals.
(a) $\frac{1}{6}$ (b) $\frac{1}{10}$ (c) $\frac{1}{25}$

3 Find the reciprocal of each of these numbers.
Give your answers as fractions or mixed numbers.
(a) $\frac{3}{5}$ (b) $\frac{4}{9}$ (c) $2\frac{2}{5}$

You may use your calculator for question **4**.

4 Find the reciprocal of each of these numbers.
Give your answers as decimals.
(a) 25   (b) 0.2   (c) 6.4

EXERCISE 29.1H

Expand these.

1 $7(3a + 6b)$
2 $5(2c + 3d)$
3 $4(3e - 5f)$
4 $3(7g - 2h)$
5 $3(4i + 2j - 3k)$
6 $3(5m - 2n + 3p)$
7 $6(4r - 3s - 2t)$
8 $8(4r + 2s + t)$
9 $4(3u + 5v)$
10 $6(4w + 3x)$
11 $2(5y + z)$
12 $4(3y + 2z)$
13 $5(3v + 2)$
14 $3(7 + 4w)$
15 $5(1 - 3a)$
16 $3(8g - 5)$

EXERCISE 29.2H

Expand the brackets and simplify these.

1 (a) $3(4a + 5) + 2(3a + 4)$
   (b) $5(4b + 3) + 3(2b + 1)$
   (c) $2(3 + 6c) + 4(5 + 7c)$
2 (a) $2(4x + 5) + 3(5x - 2)$
   (b) $4(3y + 2) + 5(3y - 2)$
   (c) $3(4 + 7z) + 2(3 - 5z)$
3 (a) $4(4s + 3t) + 5(2s + 3t)$
   (b) $3(4v + 5w) + 2(3v + 2w)$
   (c) $6(2x + 5y) + 3(4x + 2y)$
   (d) $2(5v + 4w) + 3(2v + w)$
4 (a) $5(2n + 5p) + 4(2n - 5p)$
   (b) $3(4q + 6r) + 5(2q - 3r)$
   (c) $7(3d + 2e) + 5(3d - 2e)$
   (d) $5(3f + 8g) + 4(3f - 9g)$
   (e) $4(5h - 6j) - 6(2h - 5j)$
   (f) $4(5k - 6m) - 3(2k - 5m)$

EXERCISE 29.3H

Factorise these.

1 (a) $8x + 20$
   (b) $3x + 6$
   (c) $9x - 12$
   (d) $5x - 30$
2 (a) $16 + 8x$
   (b) $9 + 15x$
   (c) $12 - 16x$
   (d) $8 - 12x$
3 (a) $4x^2 + 16x$
   (b) $6x^2 + 30x$
   (c) $8x^2 - 20x$
   (d) $9x^2 - 15x$

EXERCISE 29.4H

Expand the brackets and simplify these.

1 (a) $(a + 5)(a + 3)$
   (b) $(b + 2)(b + 4)$
   (c) $(4 + c)(3 + c)$
2 (a) $(2d + 3)(4d - 3)$
   (b) $(5e + 4)(3e - 2)$
   (c) $(3 + 8f)(2 - 5f)$
3 (a) $(3g - 2)(5g - 6)$
   (b) $(4h - 5)(3h - 7)$
   (c) $(5j - 6)(3j - 8)$
4 (a) $(3k + 7)(4k - 5)$
   (b) $(2 + 7m)(3 - 8m)$
   (c) $(4 + 3n)(2 - 5n)$
5 (a) $(2 + 5p)(3 - 7p)$
   (b) $(5r - 6)(2r - 5)$
   (c) $(3s - 2)(4s - 9)$

EXERCISE 29.5H

Simplify each of the following, writing your answer using index notation.

1 (a) $7 \times 7 \times 7 \times 7 \times 7$
   (b) $3 \times 3 \times 3 \times 3 \times 3$
   (c) $2 \times 2 \times 2 \times 2 \times 2 \times 2$

2 (a) $d \times d \times d \times d \times d \times d \times d$
   (b) $m \times m \times m \times m \times m \times m$
   (c) $t \times t \times t \times t \times t \times t \times t$

3 (a) $a \times a \times a \times a \times b \times b$
   (b) $c \times c \times c \times c \times d \times d \times d \times d \times d$
   (c) $r \times r \times r \times s \times s \times t \times t \times t \times t$

4 (a) $2x \times 3y \times 6z$
   (b) $2a \times 3b \times 4c$
   (c) $r \times 2s \times 3t \times 4s \times 5r$

EXERCISE 30.1H

Find the area of each of these triangles.

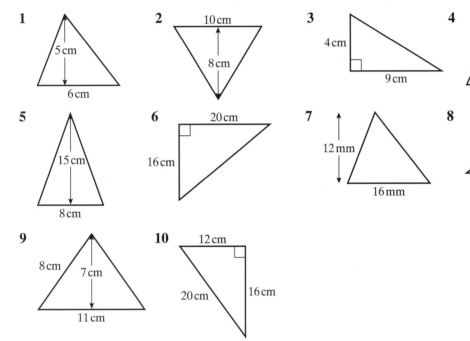

EXERCISE 30.2H

Find the area of each of these parallelograms.

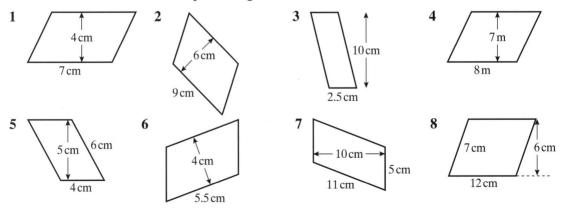

Find the missing length in each of these diagrams.

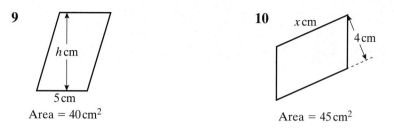

9

h cm

5 cm

Area = 40 cm^2

10

x cm

4 cm

Area = 45 cm^2

EXERCISE 30.3H

Find the size of the lettered angles. Give a reason for each answer.

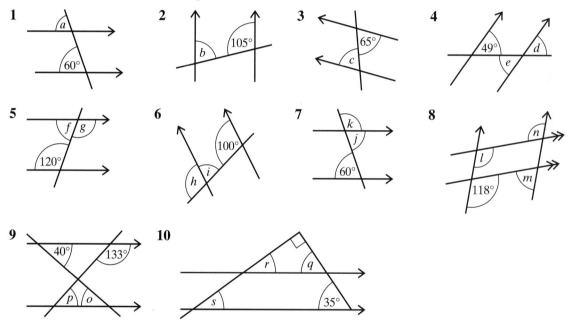

1

a

60°

2

b

105°

3

65°

c

4

49°

d

e

5

f g

120°

6

100°

h i

7

k

j

60°

8

n

l

m

118°

9

40°

133°

p o

10

r q

s

35°

EXERCISE 30.4H

Find the size of the lettered angles. Give a reason for each answer.

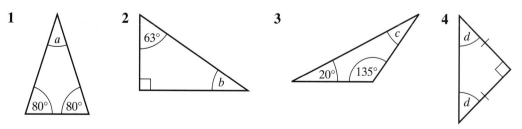

1

a

80° 80°

2

63°

b

3

c

20° 135°

4

d

d

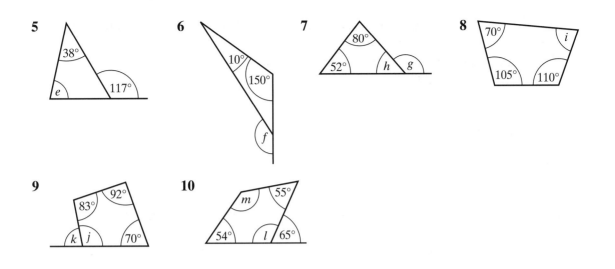

EXERCISE 30.5H

1 Name each of these quadrilaterals.

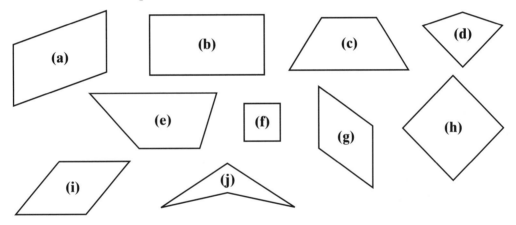

2 Name the quadrilateral or quadrilaterals which have the following properties.
   (a) Four right angles
   (b) Both pairs of opposite sides parallel
   (c) Equal diagonals
   (d) At least one pair of opposite sides parallel
   (e) Diagonals that bisect each other

3 Plot each set of points on squared paper and join them in order to make quadrilateral. Use a different grid for each part.
Write down the special name of each quadrilateral.
(a) (3, 0), (5, 2), (3, 4), (1, 2)
(b) (2, 1), (4, 1), (4, 5), (2, 5)
(c) (1, 2), (3, 1), (3, 6), (1, 7)
(d) (2, 1), (2, 5), (8, 2), (8, 4)

4 A rhombus is a special type of parallelogram.
What extra properties does a rhombus have?

5 A quadrilateral has angles of 80°, 100°, 80°, 100° in order as you work your way around the quadrilateral. The sides are not all the same length.
Which special quadrilateral could have these as their angles?
Draw the quadrilateral and mark on the angles.

EXERCISE 30.6H

1 A polygon has nine sides.
Work out the sum of the interior angles of this polygon.

2 A polygon has 13 sides.
Work out the sum of the interior angles of this polygon.

3 Four of the exterior angles of a hexagon are 93°, 50°, 37° and 85°.
The other two angles are equal.
(a) Work out the size of these equal exterior angles.
(b) Work out the interior angles of the hexagon.

4 Four of the interior angles of a pentagon are 170°, 80°, 157°, and 75°.
(a) Work out the size of the other interior angle.
(b) Work out the exterior angles of the pentagon.

5 A regular polygon has 18 sides.
Find the size of the exterior and the interior angle of this polygon.

6 A regular polygon has 24 sides.
Find the size of the exterior and the interior angle of this polygon.

7 A regular polygon has an exterior angle of 12°.
Work out the number of sides that the polygon has.

8 A regular polygon has an interior angle of 172°.
Work out the number of sides that the polygon has.

EXERCISE 31.1H

1 For each pair of fractions
- find the common denominator.
- state which is the bigger fraction.

(a) $\frac{7}{8}$ or $\frac{3}{4}$ (b) $\frac{5}{9}$ or $\frac{7}{11}$ (c) $\frac{1}{6}$ or $\frac{3}{20}$

2 Work out these.

(a) $\frac{3}{7} + \frac{2}{7}$ (b) $\frac{7}{15} + \frac{4}{15}$

(c) $\frac{8}{11} - \frac{3}{11}$ (d) $\frac{11}{17} - \frac{8}{17}$

(e) $\frac{7}{16} + \frac{3}{16}$ (f) $\frac{7}{9} + \frac{4}{9}$

(g) $\frac{7}{12} - \frac{5}{12}$ (h) $\frac{8}{11} + \frac{5}{11}$

(i) $2\frac{4}{7} + 3\frac{1}{7}$ (j) $4\frac{5}{6} - 1\frac{1}{6}$

(k) $5\frac{9}{13} - \frac{4}{13}$ (l) $4\frac{3}{8} - 1\frac{5}{8}$

3 Work out these.

(a) $\frac{2}{9} + \frac{1}{3}$ (b) $\frac{7}{12} + \frac{1}{4}$

(c) $\frac{3}{4} - \frac{1}{10}$ (d) $\frac{13}{16} - \frac{3}{8}$

(e) $\frac{7}{8} + \frac{1}{3}$ (f) $\frac{4}{5} + \frac{5}{6}$

(g) $\frac{7}{12} - \frac{1}{8}$ (h) $\frac{9}{20} + \frac{3}{4}$

(i) $\frac{7}{11} + \frac{3}{5}$ (j) $\frac{7}{12} + \frac{7}{10}$

(k) $\frac{7}{8} - \frac{1}{6}$ (l) $\frac{7}{15} - \frac{3}{20}$

4 Work out these.

(a) $4\frac{1}{4} + 3\frac{1}{3}$ (b) $6\frac{8}{9} - 1\frac{2}{3}$

(c) $5\frac{3}{8} + \frac{1}{4}$ (d) $5\frac{11}{16} - 2\frac{1}{8}$

(e) $2\frac{5}{6} + 3\frac{1}{4}$ (f) $6\frac{8}{9} - 2\frac{1}{6}$

(g) $3\frac{5}{8} + 4\frac{7}{10}$ (h) $5\frac{7}{11} - 5\frac{1}{3}$

(i) $4\frac{3}{4} + 3\frac{2}{7}$ (j) $6\frac{1}{4} - 2\frac{2}{3}$

(k) $7\frac{1}{9} - 2\frac{1}{2}$ (l) $5\frac{3}{10} - 4\frac{4}{5}$

EXERCISE 31.2H

1 Change these mixed numbers to improper fractions.

(a) $4\frac{3}{5}$ (b) $6\frac{1}{4}$ (c) $3\frac{4}{7}$ (d) $1\frac{5}{9}$

(e) $4\frac{5}{6}$ (f) $7\frac{3}{10}$ (g) $4\frac{7}{8}$

2 Work out these.
Write your answers as proper fractions or mixed numbers in their lowest terms.

(a) $\frac{3}{7} \times 5$ (b) $\frac{5}{9} \times 6$ (c) $\frac{3}{5} \div 4$

(d) $6 \times \frac{5}{11}$ (e) $\frac{2}{9} \div 4$ (f) $9 \div \frac{3}{8}$

3 Work out these.
Write your answers as proper fractions or mixed numbers in their lowest terms.

(a) $\frac{2}{3} \times \frac{5}{7}$ (b) $\frac{1}{8} \times \frac{5}{6}$

(c) $\frac{7}{9} \times \frac{2}{5}$ (d) $\frac{5}{8} \div \frac{3}{4}$

(e) $\frac{3}{8} \div \frac{1}{3}$ (f) $\frac{4}{9} \times \frac{5}{11}$

(g) $\frac{6}{7} \times \frac{1}{8}$ (h) $\frac{7}{15} \div \frac{2}{3}$

(i) $\frac{7}{12} \times \frac{3}{8}$ (j) $\frac{9}{16} \div \frac{7}{12}$

(k) $\frac{7}{10} \div \frac{5}{12}$ (l) $\frac{7}{30} \times \frac{10}{21}$

4 Work out these.
Write your answers as proper fractions or mixed numbers in their lowest terms.

(a) $4\frac{3}{4} \times 1\frac{7}{9}$ (b) $3\frac{2}{3} \times \frac{1}{5}$

(c) $4\frac{2}{5} \div 2\frac{4}{5}$ (d) $1\frac{3}{11} \div 3\frac{1}{2}$

(e) $4\frac{1}{2} \times 3\frac{2}{3}$ (f) $3\frac{5}{9} \div 2\frac{2}{3}$

(g) $3\frac{2}{7} \times 1\frac{5}{9}$ (h) $2\frac{5}{8} \div 1\frac{5}{6}$

(i) $1\frac{7}{15} \times 12\frac{1}{2}$ (j) $5\frac{3}{5} \div 1\frac{3}{4}$

(k) $6\frac{2}{9} \times 2\frac{1}{8}$ (l) $7\frac{1}{2} \div 2\frac{3}{5}$

EXERCISE 31.3H

1 Work out these.

(a) $\frac{3}{4} + \frac{1}{6}$ (b) $\frac{5}{8} - \frac{2}{7}$

(c) $\frac{5}{9} \times \frac{3}{8}$ (d) $\frac{7}{16} \div \frac{5}{12}$

(e) $1\frac{4}{5} + 2\frac{3}{4}$ (f) $6\frac{3}{7} - 2\frac{1}{3}$

(g) $5\frac{3}{5} \times 4$ (h) $4\frac{5}{9} \div 1\frac{1}{6}$

2 Write these fractions in their lowest terms.

(a) $\frac{40}{125}$ (b) $\frac{28}{49}$ (c) $\frac{72}{192}$

(d) $\frac{225}{350}$ (e) $\frac{17}{153}$

3 Write these improper fractions as mixed numbers.

(a) $\frac{120}{72}$ (b) $\frac{150}{13}$ (c) $\frac{86}{19}$

(d) $\frac{192}{54}$ (e) $\frac{302}{17}$

4 Calculate
(a) the perimeter of this rectangle.
(b) the area of this rectangle.

$5\frac{3}{5}$ cm

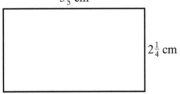

$2\frac{1}{4}$ cm

EXERCISE 31.4H

1 Change each of these fractions to a decimal.
If necessary, give your answer to 3 decimal places.

(a) $\frac{7}{8}$ (b) $\frac{7}{100}$

(c) $\frac{5}{9}$ (d) $\frac{2}{11}$

2 State whether each of these fractions gives a recurring or a terminating decimal.
Give your reasons.

(a) $\frac{3}{4}$ (b) $\frac{5}{6}$ (c) $\frac{5}{11}$

(d) $\frac{2}{25}$ (e) $\frac{7}{32}$

3 (a) Find the recurring decimal equivalent to $\frac{3}{101}$.
(b) How many digits are there in the repeating pattern?

EXERCISE 31.5H

Work out these. As far as possible, write down only your final answer.

1 $3.4 + 6.1$ **2** $4.3 + 3.6$

3 $5.8 - 2.3$ **4** $7.9 - 4.4$

5 $3.7 + 2.6$ **6** $5.8 + 3.4$

7 $7.2 - 0.9$ **8** $5.4 - 3.5$

9 $8.6 + 2.7$ **10** $6.9 + 4.6$

11 $6.7 - 5.8$ **12** $6.3 - 2.9$

EXERCISE 31.6H

1 Work out these.

(a) 5×0.4 (b) 0.6×8

(c) 4×0.7 (d) 0.9×6

(e) 0.7×0.3 (f) 0.9×0.4

(g) 50×0.7 (h) 0.4×80

(i) 0.7×0.1 (j) 0.4×0.2

(k) $(0.8)^2$ (l) $(0.2)^2$

2 Work out these.

(a) $6 \div 0.3$ (b) $4.8 \div 0.2$

(c) $2.4 \div 0.6$ (d) $7.2 \div 0.4$

(e) $33 \div 1.1$ (f) $60 \div 1.5$

(g) $12 \div 0.4$ (h) $35 \div 0.7$

(i) $64 \div 0.8$ (j) $32 \div 0.2$

(k) $2.17 \div 0.7$ (l) $47.5 \div 0.5$

3 Work out these.
(a) 3.6×1.4 (b) 5.8×2.6
(c) 8.1×4.3 (d) 6.5×3.2
(e) 74×1.7 (f) 64×3.8
(g) 2.9×7.6 (h) 11.4×3.2
(i) 25.2×0.8 (j) 2.67×0.9
(k) 8.45×1.2 (l) 7.26×2.4

4 Work out these.
(a) $23.6 \div 0.4$ (b) $23.4 \div 0.8$
(c) $18.2 \div 0.7$ (d) $31.2 \div 0.6$
(e) $42.3 \div 0.9$ (f) $75.6 \div 1.2$
(g) $5.28 \div 0.3$ (h) $7.56 \div 0.7$
(i) $63.2 \div 0.2$ (j) $6.27 \div 1.1$
(k) $3.51 \div 1.3$ (l) $8.19 \div 1.3$

EXERCISE 31.7H

1 Write down the multiplier that will increase an amount by
(a) 17%.   (b) 30%.   (c) 73%.
(d) 6%.   (e) 1%.   (f) 12.5%.
(g) 160%.

2 Write down the multiplier that will decrease an amount by
(a) 13%.   (b) 40%.   (c) 35%.
(d) 8%.   (e) 4%.   (f) 27%.
(g) 15.5%.

3 Mrs Green bought an antique for £200.
She later sold it at 250% profit.
What did she sell it for?

4 Jane earns £14 500 per year.
She receives an increase of 2%.
Find her new salary.

5 In a sale all items are reduced by 20%.
Shamir bought a computer in the sale.
The original price was £490.
What was the sale price?

6 Graham invested £3500 at 4% compound interest.
What was the investment worth at the end of 5 years?
Give your answer to the nearest pound.

7 A car decreased in value by 11% per year.
If it cost £16 500 new, what was it worth after 4 years?
Give your answer to a suitable degree of accuracy.

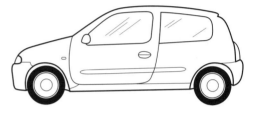

8 In a certain country the population rose by 5% every year from 1999 to 2004.
If the population was 26.5 million in 1999, what was the population in 2004?
Give your answer in millions to the nearest 0.1 of a million.

9 Jane invested £4500 with compound interest for 3 years.
She could receive either 3% interest every 6 months or 6% interest every year.
Which should Jane choose.
How much more will she receive?

10 Prices went up by 2% in 2003, 3% in 2004 and 2.5% in 2005.
If an item cost £32 at the start of 2003 what would it cost at the end of 2005?

EXERCISE 32.1H

1 For a project, Rebecca recorded the ages of 100 cars as they passed the school gates one morning. Here are her results.

Age (a years)	$0 \leqslant a < 2$	$2 \leqslant a < 4$	$4 \leqslant a < 6$	$6 \leqslant a < 8$	$8 \leqslant a < 10$	$10 \leqslant a < 12$	$12 \leqslant a < 14$
Frequency	16	23	24	17	12	7	1

  (a) Draw a frequency diagram to show these data.
  (b) Which of the intervals is the modal group?

2 The manager of a leisure centre recorded the weights of 120 men. Here are the results.

Weight (w kg)	$60 \leqslant w < 65$	$65 \leqslant w < 70$	$70 \leqslant w < 75$	$75 \leqslant w < 80$	$80 \leqslant w < 85$	$85 \leqslant w < 90$
Frequency	4	18	36	50	10	2

  (a) Draw a frequency diagram to represent these data.
  (b) Which of the intervals is the modal group?
  (c) Which of the intervals contains the median value?

3 This frequency diagram shows the times taken by a group of girls to run a race.
  (a) How many girls took longer than 9 minutes?
  (b) How many girls took part in the race?
  (c) What percentage of the girls took less than 7 minutes?
  (d) What is the modal finishing time?
  (e) Use the diagram to draw up a grouped frequency table like those in questions **1** and **2**.

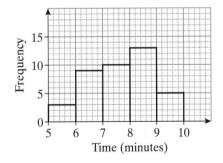

EXERCISE 32.2H

1 The table shows the heights of 40 plants.

Height (h cm)	$3 \leqslant h < 4$	$4 \leqslant h < 5$	$5 \leqslant h < 6$	$6 \leqslant h < 7$	$7 \leqslant h < 8$	$8 \leqslant h < 9$
Frequency	1	7	10	12	8	2

Draw a frequency polygon to show these data.

2 The table shows the time taken for a group of children to get from home to school.

Time (t mins)	$0 \leqslant t < 5$	$5 \leqslant t < 10$	$10 \leqslant t < 15$	$15 \leqslant t < 20$	$20 \leqslant t < 25$	$25 \leqslant t < 30$
Frequency	3	15	27	34	19	2

Draw a frequency polygon to show these data.

3 The ages of all of the people under 70 in a small village were recorded in 1985 and 2005. The results are given in the table below.

Age (a years)	$0 \leqslant a < 10$	$10 \leqslant a < 20$	$20 \leqslant a < 30$	$30 \leqslant a < 40$	$40 \leqslant a < 50$	$50 \leqslant a < 60$	$60 \leqslant a < 70$
Frequency 1985	85	78	70	53	40	28	18
Frequency 2005	50	51	78	76	62	64	56

(a) On the same grid, draw a frequency polygon for each year.

(b) Use the diagram to compare the distribution of ages in the two years.

EXERCISE 32.3H

1 Bill grows tomatoes. As an experiment he divided his land into eight plots.
He used a different amount of fertiliser on each plot.
The table shows the weight of tomatoes he got from each of the plots.

Amount of fertiliser (g/m²)	10	20	30	40	50	60	70	80
Weight of tomatoes (kg)	36	41	58	60	70	76	75	92

(a) Draw a scatter diagram to show this information.
(b) Describe the correlation shown in the scatter diagram.
(c) The mean of the amount of fertiliser is 45 g/m². Calculate the mean weight of the tomatoes.
(d) Plot the point that has these means as coordinates.
(e) Draw a line of best fit on your scatter diagram.
(f) What weight of tomatoes should Bill expect to get if he used 75 g/m²?

2 The table shows the prices and mileages of seven second-hand cars of the same model.

Price (£)	6000	3500	1000	8500	5500	3500	7000
Mileage	29 000	69 000	92 000	17 000	53 000	82 000	43 000

(a) Draw a scatter diagram to show this information.
(b) Describe the correlation shown in the scatter diagram.
(c) The mean price (£) is £5000 and the mean of the mileages is 55 000.
(d) Plot the point that has these means as coordinates.
(e) Draw a line of best fit on your scatter diagram.
(f) Use your line of best fit to estimate
   (i) the price of this model of car which has covered 18 000 miles.
   (ii) the mileage of this model of car which costs £4000.

3 The heights of 10 daughters, all aged 20, and their fathers are given in the table below.

Height of father (cm)	167	168	169	171	172	172	174	175	176	182
Height of daughter (cm)	164	166	166	168	169	170	170	171	173	177

(a) Draw a scatter diagram to show this information.
(b) Describe the correlation shown in the scatter diagram.
(c) The mean height of the fathers 172.6 cm. Calculate the mean height of the daughters.
(d) Plot the point that has these means as coordinates.
(e) Draw a line of best fit on your scatter diagram.
(f) Use your line of best fit to estimate the height of a 20-year-old daughter whose father is 180 cm tall.

EXERCISE 33.1H

Find the circumferences of circles with these diameters.

1	8 cm	**2**	17 cm	**3**	39.2 cm
4	116 mm	**5**	5.1 m	**6**	6.32 m
7	14 cm	**8**	23 cm	**9**	78 mm
10	39 mm	**11**	4.4 m	**12**	2.75 m

EXERCISE 33.2H

1 Find the areas of circles with the following radii.

(a)	17 cm	**(b)**	23 cm
(c)	67 cm	**(d)**	43 mm
(e)	74 mm	**(f)**	32 cm
(g)	58 cm	**(h)**	4.3 cm
(i)	8.7 cm	**(j)**	47 m
(k)	1.9 m	**(l)**	2.58 m

2 Find the areas of circles with the following diameters.

(a)	18 cm	**(b)**	28 cm
(c)	68 cm	**(d)**	38 mm
(e)	78 mm	**(f)**	58 cm
(g)	46 cm	**(h)**	6.4 cm
(i)	7.6 cm	**(j)**	32 m
(k)	3.4 m	**(l)**	4.32 m

EXERCISE 33.3H

Find the area of each of these shapes. Break them down into rectangles and right-angled triangles first.

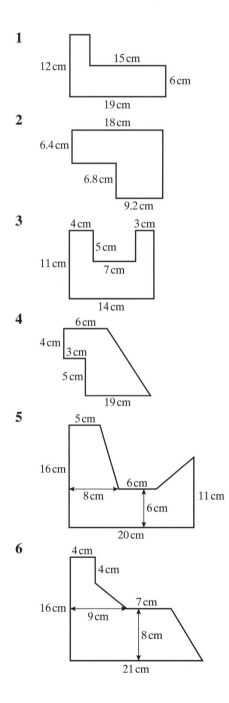

EXERCISE 33.4H

Find the volume of each of these shapes.

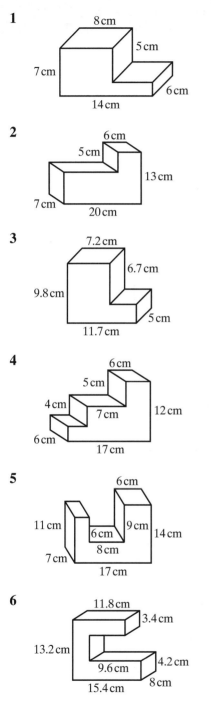

1

8 cm
5 cm
7 cm
14 cm
6 cm

2

6 cm
5 cm
13 cm
7 cm
20 cm

3

7.2 cm
6.7 cm
9.8 cm
5 cm
11.7 cm

4

6 cm
5 cm
4 cm
7 cm
12 cm
6 cm
17 cm

5

6 cm
11 cm
6 cm
9 cm
14 cm
7 cm
8 cm
17 cm

6

11.8 cm
3.4 cm
13.2 cm
4.2 cm
9.6 cm
8 cm
15.4 cm

EXERCISE 33.5H

Find the volume of each of these prisms.

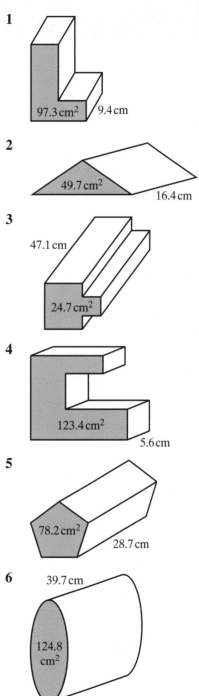

1

97.3 cm² 9.4 cm

2

49.7 cm²
16.4 cm

3

47.1 cm
24.7 cm²

4

123.4 cm²
5.6 cm

5

78.2 cm²
28.7 cm

6

39.7 cm
124.8 cm²

EXERCISE 33.6H

Find the volumes of cylinders with these dimensions.

1 Radius 7 cm and height 29 cm
2 Radius 13 cm and height 27 cm
3 Radius 25 cm and height 80 cm
4 Radius 14 mm and height 35 mm
5 Radius 28 mm and height 8 mm
6 Radius 0.6 mm and height 5.1 mm
7 Radius 1.7 m and height 5 m
8 Radius 2.6 m and height 3.4 m

EXERCISE 33.7H

Find the curved surface areas of cylinders with these dimensions.

1 Radius 9 cm and height 16 cm
2 Radius 13 cm and height 21 cm
3 Radius 27 cm and height 12 cm
4 Radius 17 mm and height 35 mm
5 Radius 12 mm and height 6 mm
6 Radius 3.7 mm and height 63 mm
7 Radius 1.9 m and height 19 m
8 Radius 2.7 m and height 4.3 m

Find the total surface areas of cylinders with these dimensions.

9 Radius 8 cm and height 11 cm
10 Radius 17 cm and height 28 cm
11 Radius 29 cm and height 15 cm
12 Radius 32 mm and height 8 mm
13 Radius 35 mm and height 12 mm
14 Radius 3.9 mm and height 45 mm
15 Radius 0.8 m and height 7 m
16 Radius 2.9 m and height 1.7 m

EXERCISE 33.8H

Draw the plan view, front elevation and side elevation of each of these objects.

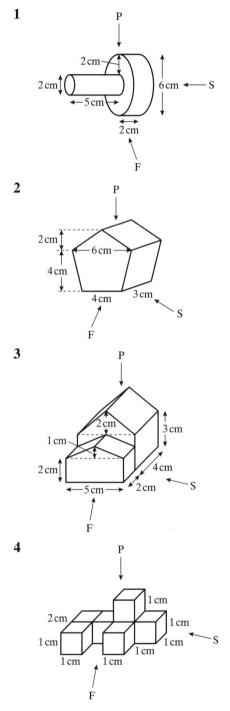

EXERCISE 34.1H

Solve these equations.

1 $2x - 3 = 7$

2 $2x + 2 = 8$

3 $2x - 9 = 3$

4 $3x - 2 = 7$

5 $6x + 2 = 26$

6 $3x + 2 = 17$

7 $4x - 5 = 3$

8 $4x + 2 = 8$

9 $2x - 7 = 10$

10 $5x + 12 = 7$

11 $x^2 + 3 = 19$

12 $x^2 - 2 = 7$

13 $y^2 - 1 = 80$

14 $11 - 3x = 2$

15 $4x - 12 = -18$

EXERCISE 34.2H

Solve these equations.

1 $3(x - 2) = 18$

2 $2(1 + x) = 8$

3 $3(x - 5) = 6$

4 $2(x + 3) = 10$

5 $5(x - 2) = 15$

6 $2(x + 3) = 10$

7 $5(x - 4) = 20$

8 $4(x + 1) = 16$

9 $2(x - 7) = 8$

10 $3(2x + 3) = 18$

11 $5(2x - 3) = 15$

12 $2(3x - 2) = 14$

13 $5(2x - 3) = 40$

14 $4(x - 3) = 6$

15 $2(2x - 3) = 8$

EXERCISE 34.3H

Solve these equations.

1 $5x - 1 = 3x + 5$

2 $5x + 1 = 2x + 13$

3 $7x - 2 = 2x + 8$

4 $6x + 1 = 4x + 21$

5 $9x - 10 = 4x + 5$

6 $5x - 8 = 3x - 6$

7 $6x + 2 = 10 - 2x$

8 $2x - 10 = 5 - 3x$

9 $15 + 3x = 2x + 18$

10 $2x - 5 = 4 - x$

11 $3x - 2 = x + 7$

12 $x - 1 = 2x - 6$

13 $2x - 4 = 2 - x$

14 $9 - x = x + 5$

15 $3x - 2 = x - 8$

EXERCISE 34.4H

Solve these equations.

1 $\dfrac{x}{2} = 7$

2 $\dfrac{x}{5} - 2 = 1$

3 $\dfrac{x}{4} + 5 = 8$

4 $\dfrac{x}{3} - 5 = 5$

5 $\dfrac{x}{6} + 3 = 4$

6 $\dfrac{x}{5} + 1 = 4$

7 $\dfrac{x}{8} - 3 = 9$

8 $\dfrac{x}{4} + 1 = 3$

9 $\dfrac{x}{7} + 5 = 6$

10 $\dfrac{x}{4} + 5 = 4$

EXERCISE 34.5H

For each of questions **1** to **5**, solve the inequality and show the solution on a number line.

1 $x - 2 > 1$

2 $x + 1 < 3$

3 $3x - 2 \geqslant 7$

4 $2x + 1 \leqslant 6$

5 $3x - 6 \geqslant 0$

For each of questions **6** to **15**, solve the inequality.

6 $7 \leqslant 2x - 1$

7 $5x < x + 12$

8 $4x \geqslant x + 9$

9 $4 + x < 0$

10 $3x + 1 \leqslant 2x + 6$

11 $2(x - 3) > x$

12 $5(x + 1) > 3x + 10$

13 $7x + 5 \leqslant 2x + 30$

14 $5x + 2 < 7x - 4$

15 $3(3x + 2) \geqslant 2(x + 10)$

EXERCISE 35.1H

1 Write each of these ratios in its lowest terms.
   (a) 8 : 6
   (b) 20 : 50
   (c) 35 : 55
   (d) 8 : 24 : 32
   (e) 15 : 25 : 20

2 Write each of these ratios in its lowest terms.
   (a) 200 g : 500 g
   (b) 60p : £3
   (c) 1 minute : 25 seconds
   (d) 2 m : 80 cm
   (e) 500 g : 3 kg

3 A bar of brass contains 400 g of copper and 200 g of zinc.
   Write the ratio of copper to zinc in its lowest terms.

4 Teri, Jannae and Abi receive £200, £350 and £450 respectively as their dividends in a joint investment.
   Write the ratio of their dividends in its lowest terms.

5 Three saucepans hold 500 ml, 1 litre and 2.5 litres respectively.
   Write the ratio of their capacities in its lowest terms.

EXERCISE 35.2H

1 Write each of these ratios in the form 1 : n.
   (a) 2 : 10   (b) 5 : 30
   (c) 2 : 9   (d) 4 : 9
   (e) 50 g : 30 g   (f) 15p : £3
   (g) 25 cm : 6 m   (h) 20 : 7
   (i) 4 mm : 1 km

2 On a map a distance of 12 mm represents a distance of 3 km.
   What is the scale of the map in the form 1 : n?

3 A picture is enlarged on a photocopier from 25 mm wide to 15 cm wide.
   What is the ratio of the picture to the enlargement in the form 1 : n?

EXERCISE 35.3H

1 A photo is enlarged in the ratio 1 : 5.
   (a) The length of the small photo is 15 cm.
       What is the length of the large photo?
   (b) The width of the large photo is 45 cm.
       What is the width of the small photo?

2 To make a dressing for her lawn,
Rupinder mixes loam and sand in the
ratio 1 : 3.
  (a) How much sand should she mix
      with two buckets of loam?
  (b) How much loam should she mix
      with 15 buckets of sand?

3 To make mortar, Fred mixes 1 part
cement with 5 parts sand.
  (a) How much sand does he mix with
      500 g of cement?
  (b) How much cement does he mix
      with 4.5 kg of sand?

4 A rectangular picture is 6 cm wide.
It is enlarged in the ratio 1 : 4.
How wide is the enlargement?

5 The Michelin motoring atlas of France
has a scale of 1 cm to 2 km.
  (a) On the map the distance between
      Metz and Nancy is 25 cm.
      How far is the actual distance
      between the two towns?
  (b) The actual distance between Caen
      and Falaise is 33 km.
      How far is this on the map?

6 Graham is making pastry.
To make enough for five people he
uses 300 g of flour.
How much flour should he use for
eight people?

7 To make a solution of a chemical a
scientist mixes 2 parts chemical with
25 parts water.
  (a) How much water should he mix
      with 10 ml of chemical?
  (b) How much chemical should he
      mix with 1 litre of water?

8 The ratio of the sides of two rectangles
is 2 : 5.
  (a) The length of the small rectangle is
      4 cm.
      How long is the big rectangle?
  (b) The width of the big rectangle is
      7.5 cm.
      How wide is the small rectangle?

9 Jason mixes 3 parts black paint with
4 parts white paint to make dark grey
paint.
  (a) How much white paint does he
      mix with 150 ml of black paint?
  (b) How much black paint does he
      mix with 1 litre of white paint?

10 In an election the number of votes was
shared between the Labour,
Conservatives and other parties in the
ratio 5 : 4 : 2.
Labour received 7500 votes.
  (a) How many votes did the
      Conservatives receive?
  (b) How many votes did the other
      parties receive?

EXERCISE 35.4H

Do not use your calculator for questions
1 to 5.

1 Share £40 between Paula and Tariq in
the ratio 3 : 5.

2 Paint is mixed in the ratio 2 parts
black paint to 3 parts white paint to
make 10 litres of grey paint.
  (a) How much black paint is used?
  (b) How much white paint is used?

3 A metal alloy is made up of copper, iron and nickel in the ratio 3 : 4 : 2. How much of each metal is there in 450 g of the alloy?

4 Inderjit worked 6 hours one day. The time he spent on filing, writing and computing is in the ratio 2 : 3 : 7. How long did he spend computing?

5 Daisy and Emily invested £5000 and £8000 respectively in a business venture.
They agreed to share the profits in the ratio of their investment.
Emily received £320.
What was the total profit?

You may use your calculator for questions **6** to **8**.

6 Shahida spends her pocket money on sweets, magazines and clothes in the ratio 2 : 3 : 7.
She receives £15 a week.
How much does she spend on sweets?

7 In a questionnaire the three possible answers are 'Yes', 'No' and 'Don't know'.
The answers from a group of 456 people are in the ratio 10 : 6 : 3.
How many 'Don't knows' are there?

8 Iain and Stephen bought a house between them in Spain.
Iain paid 60% of the cost and Stephen 40%.
(a) Write the ratio of the amounts they paid in its lowest terms.
(b) The house cost 210 000 euros. How much did each pay?

EXERCISE 35.5H

1 An 80 g bag of Munchoes costs 99p and a 200 g bag of Munchoes costs £2.19.
Show which is the better value.

2 Baxter's lemonade is sold in 2 litre bottles for £1.29 and in 3 litre bottles for £1.99.
Show which is the better value.

3 Butter is sold in 200 g tubs for 95p and in 450 g packets for £2.10.
Show which is the better value.

4 Fruit yogurt is sold in packs of 4 tubs for 79p and in packs of 12 tubs for £2.19.
Show which is the better value.

5 There are two packs of minced meat on the reduced price shelf at the supermarket, a 1.8 kg pack reduced to £2.50 and a 1.5 kg pack reduced to £2.
Show which is the better value.

6 Smoothie shaving gel costs £1.19 for the 75 ml bottle and £2.89 for the 200 ml bottle.
Show which is better value.

7 A supermarket sells cans of cola in two different sized packs: a pack of 12 cans costs £4.30 and a pack of 20 cans costs £7.25.
Which pack gives the better value?

8 Sudso washing powder is sold in 3 sizes: 750 g for £3.15, 1.5 kg for £5.99 and 2.5 kg for £6.99.
Which packet gives the best value?

EXERCISE 36.1H

1 For each of these sets of data
(i) find the mode. (ii) find the median. (iii) find the range. (iv) calculate the mean.

(a)

Score on biased dice	Number of times thrown
1	52
2	46
3	70
4	54
5	36
6	42
Total	300

(b)

Number of drawing pins in a box	Number of boxes
98	5
99	14
100	36
101	28
102	17
103	13
104	7
Total	120

(c)

Number of snacks per day	Frequency
0	23
1	68
2	39
3	21
4	10
5	3
6	1

(d)

Number of letters received on Monday	Frequency
0	19
1	37
2	18
3	24
4	12
5	5
6	2
7	3

2 Gift tokens cost £1, £5, £10, £20 or £50 each.

The frequency table below shows the numbers of each value of gift token sold in one bookstore on a Saturday.

Calculate the mean value of gift token bought in the bookstore that Saturday.

Price of gift token (£)	1	5	10	20	50
Number of tokens sold	12	34	26	9	1

3 A sample of people were asked how many visits to the cinema they had made in one month.

None of those asked had made more than eight visits to the cinema.

The table below shows the data.

Calculate the mean number of visits to the cinema.

Number of visits	0	1	2	3	4	5	6	7	8
Frequency	136	123	72	41	18	0	5	1	4

EXERCISE 36.2H

1 For each of these sets of data, calculate an estimate of

(**i**) the range.

(**ii**) the mean.

(**a**)

Number of trains arriving late each day (x)	Number of days (f)
0–4	19
5–9	9
10–14	3
15–19	0
20–24	1
Total	32

(**b**)

Number of weeds per square metre (x)	Number of square metres (f)
0–14	204
15–29	101
30–44	39
45–59	13
60–74	6
75–89	2

(c)

Number of books sold (x)	Frequency (f)
60–64	3
65–69	12
70–74	23
75–79	9
80–84	4
85–89	1

(d)

Number of days absent (x)	Frequency (f)
0–3	13
4–7	18
8–11	9
12–15	4
16–19	0
20–23	1
24–27	3

2 The table below gives the number of sentences per chapter in a book.

Number of sentences (x)	$100 \leqslant x < 125$	$125 \leqslant x < 150$	$150 \leqslant x < 175$	$175 \leqslant x < 200$	$200 \leqslant x < 225$
Frequency	1	9	8	5	2

  (a) What is the modal group?
  (b) In which group is the median number of sentences?
  (c) Calculate an estimate of the mean number of sentences.

3 A group of students were asked to estimate the number of beans in a jar.
The results of their estimates are summarised in the table.
Calculate an estimate of the mean number of beans estimated by these students.

Estimated number of beans (x)	Frequency (f)
300–324	9
325–349	26
350–374	52
375–399	64
400–424	83
425–449	57
450–474	18
475–499	5

EXERCISE 36.3H

1 For each of these sets of data, calculate an estimate of
   (i) the range.   (ii) the mean.

(a)

Height of sunflower in centimetres (x)	Number of plants (f)
$100 \leqslant x < 110$	6
$110 \leqslant x < 120$	13
$120 \leqslant x < 130$	35
$130 \leqslant x < 140$	29
$140 \leqslant x < 150$	16
$150 \leqslant x < 160$	11
Total	110

(b)

Weight of egg in grams (x)	Number of eggs (f)
$20 \leqslant x < 25$	9
$25 \leqslant x < 30$	16
$30 \leqslant x < 35$	33
$35 \leqslant x < 40$	48
$40 \leqslant x < 45$	29
$45 \leqslant x < 50$	15
Total	150

(c)

Length of green bean in millimetres (x)	Frequency (f)
$60 \leqslant x < 80$	12
$80 \leqslant x < 100$	21
$100 \leqslant x < 120$	46
$120 \leqslant x < 140$	27
$140 \leqslant x < 160$	14
Total	120

(d)

Time to complete race in minutes (x)	Frequency (f)
$54 \leqslant x < 56$	1
$56 \leqslant x < 58$	4
$58 \leqslant x < 60$	11
$60 \leqslant x < 62$	6
$62 \leqslant x < 64$	2
$64 \leqslant x < 66$	1
Total	25

2 For each of these sets of data
   (i) write down the modal class.   (ii) calculate an estimate of the mean.

(a)

Height of shrub in metres (x)	Number of shrubs (f)
$0.3 \leqslant x < 0.6$	57
$0.6 \leqslant x < 0.9$	41
$0.9 \leqslant x < 1.2$	36
$1.2 \leqslant x < 1.5$	24
$1.5 \leqslant x < 1.8$	15

(b)

Weight of plum in grams (x)	Number of plums (f)
$20 \leqslant x < 30$	6
$30 \leqslant x < 40$	19
$40 \leqslant x < 50$	58
$50 \leqslant x < 60$	15
$60 \leqslant x < 70$	4

(c)

Length of journey in minutes (x)	Frequency (f)
$20 \leqslant x < 22$	6
$22 \leqslant x < 24$	20
$24 \leqslant x < 26$	38
$26 \leqslant x < 28$	47
$28 \leqslant x < 30$	16
$30 \leqslant x < 32$	3

(d)

Speed of car in miles per hour (x)	Frequency (f)
$25 \leqslant x < 30$	4
$30 \leqslant x < 35$	29
$35 \leqslant x < 40$	33
$40 \leqslant x < 45$	6
$45 \leqslant x < 50$	2
$50 \leqslant x < 55$	1

3 The table shows the monthly wages of the workers in an office.

Wages in £ (x)	$500 \leqslant x < 1000$	$1000 \leqslant x < 1500$	$1500 \leqslant x < 2000$	$2000 \leqslant x < 2500$
Frequency (f)	3	14	18	5

(a) What is the modal class?

(b) In which class is the median wage?

(c) Calculate an estimate of the mean wage.

4 The table shows the length, in seconds, of 100 calls made from a mobile phone.

Length of call in seconds (x)	$0 \leqslant x < 30$	$30 \leqslant x < 60$	$60 \leqslant x < 90$	$90 \leqslant x < 120$	$120 \leqslant x < 150$
Frequency (f)	51	25	13	7	4

Calculate an estimate of the mean length of a call.

5 The table shows the prices paid for greetings cards sold in one day by a card shop.

Calculate an estimate of the mean price, in pence, paid for a greetings card that day.

Price of greetings card in pence (x)	Frequency (f)
$75 \leqslant x < 100$	23
$100 \leqslant x < 125$	31
$125 \leqslant x < 150$	72
$150 \leqslant x < 175$	59
$175 \leqslant x < 200$	34
$200 \leqslant x < 225$	11
$225 \leqslant x < 250$	5

EXERCISE 37.1H

For each of these diagrams, find the area of the third square.

1

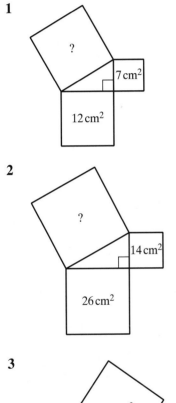

2

3

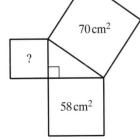

4

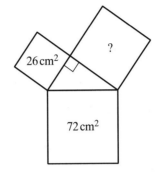

EXERCISE 37.2H

1 For each of these triangles, find the length marked x.
Where the answer is not exact, give your answer correct to 2 decimal places.

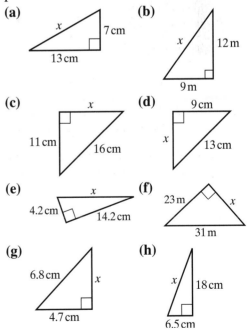

2

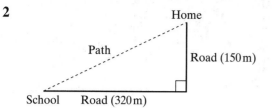

Ann can walk home from school along two roads or along a path across a field.
How much shorter is her journey if she takes the path across the field?

3

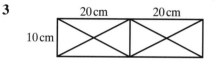

This network is made of wire.
What is the total length of wire?

4 A tangent is drawn from a point P to meet the circle, centre O, at the point T such that TP = 12.8 and PTO is a right angle. Given that the distance OP = 16.5 cm, calculate the radius of the circle.

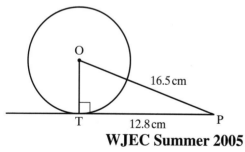

WJEC Summer 2005

5 A prism has a uniform cross-section in the shape of a triangle ABC, right angled at B and in which AC = 5.6 cm and AB = 3.4 cm. The length of the prism is 14.5 cm. Calculate the volume of the prism.

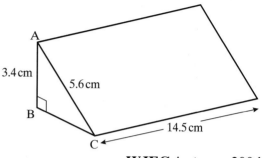

WJEC Autumn 2004

EXERCISE 37.3H

Work out whether or not each of these triangles is right-angled.
Show your working.

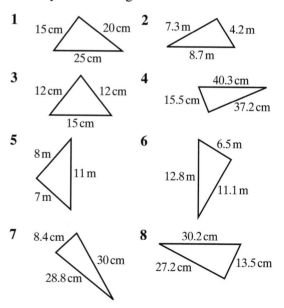

EXERCISE 37.4H

1 Find the coordinates of the midpoint
   of each of the lines in the diagram.

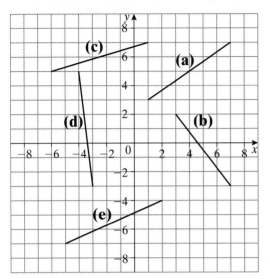

2 Find the coordinates of the midpoint
   of the line joining each of these pairs
   of points.
   Try to do them without plotting the
   points.
   (a) A(3, 7) and B(−5, 7)
   (b) C(2, 1) and D(8, 5)
   (c) E(3, 7) and F(8, 2)
   (d) G(1, 6) and H(9, 3)
   (e) I(−7, 1) and J(3, 6)
   (f) K(−5, −6) and L(−7, −3)

38 → MENTAL METHODS 2

EXERCISE 38.1H

Work these out mentally. As far as possible, write down only the final answer.

1 (a) $8 + 25$ (b) $13 + 49$
(c) $0.6 + 5.2$ (d) $142 + 59$
(e) $187 + 25$ (f) $5.8 + 12.6$
(g) $326 + 9.3$ (h) $456 + 83$
(i) $1290 + 41$ (j) $8600 + 570$

2 (a) $12 - 5$ (b) $36 - 9$
(c) $10 - 0.4$ (d) $56 - 45$
(e) $82 - 39$ (f) $141 - 27$
(g) $1.2 - 0.7$ (h) $186 - 19$
(i) $307 - 81$ (j) $1200 - 153$

3 (a) 7×9 (b) 12×8
(c) 22×7 (d) 11×12
(e) 0.4×100 (f) 19×7
(g) 48×5 (h) 25×8
(i) 63×4 (j) 42×21

4 (a) $42 \div 7$ (b) $72 \div 12$
(c) $1500 \div 3$ (d) $184 \div 8$
(e) $240 \div 20$ (f) $108 \div 18$
(g) $96 \div 16$ (h) $16 \div 100$
(i) $24 \div 0.3$ (j) $3.6 \div 0.9$

5 (a) $7 + (-3)$ (b) $-2 + 6$
(c) $-5 + 7$ (d) $-4 + (-8)$
(e) $4 + (-9)$ (f) $-5 - 1$
(g) $6 - (-2)$ (h) $12 - (-8)$
(i) $-5 - (-10)$ (j) $-7 - (-3)$

6 (a) 12×-2 (b) -4×5
(c) -6×-3 (d) -10×-4
(e) 4×-100 (f) $10 \div -5$
(g) $-6 \div 2$ (h) $-20 \div -4$
(i) $-35 \div -5$ (j) $32 \div -4$

7 Write down the square of each of these numbers.
(a) 7 (b) 9
(c) 12 (d) 14
(e) 1 (f) 30
(g) 200 (h) 0.5
(i) 0.8 (j) 0.2

8 Write down the square root of each of these numbers.
(a) 36 (b) 4 (c) 64
(d) 121 (e) 144

9 Write down the cube of each of these numbers.
(a) 3 (b) 4 (c) 10
(d) 50 (e) 0.2

10 A rectangle has sides 4.6 cm and 5.0 cm. Work out
(a) the perimeter of the rectangle.
(b) the area of the rectangle.

11 Ayeesha spends £17.81. How much change does she get from £50?

12 A square has area 121 cm². How long is its side?

13 Find 2% of £540.

14 Find two numbers the difference
between which is 4 and the product 45.

15 Deepa has a 1 kg piece of cheese.
She eats 432 g of it.
How much is left?

EXERCISE 38.2H

1 Round each of these numbers to
1 significant figure.
(a) 14.3   (b) 38
(c) 6.54   (d) 308
(e) 1210   (f) 0.78
(g) 0.61   (h) 0.053
(i) 2413.5   (j) 0.0097

2 Round each of these numbers to
1 significant figure.
(a) 8.4   (b) 18.36
(c) 725   (d) 8032
(e) 98.3   (f) 0.71
(g) 0.0052   (h) 0.019
(i) 407.511   (j) 23 095

3 Round each of these numbers to
2 significant figures.
(a) 28.7   (b) 149.3
(c) 7832   (d) 46 820
(e) 21.36   (f) 0.194
(g) 0.0489   (h) 0.003 61
(i) 0.0508   (j) 0.904

4 Round each of these numbers to
3 significant figures.
(a) 7.385   (b) 24.81
(c) 28 462   (d) 308.61
(e) 16 418   (f) 3.917
(g) 60.72   (h) 0.9135
(i) 0.004 162   (j) 2.236 06

For questions 5 to 12, round the numbers
in your calculations to 1 significant figure.
Show your working.

5 Ashad bought 24 chocolate bars at 32p
each. Estimate how much he spent.

6 Ramy earns £382 per week.
Estimate his earnings in a year.

7 Mary drove 215 miles in 3 hours
48 minutes.
Estimate her average speed.

8 A rectangle has length 9.2 cm and area
44.16 cm^2. Estimate its width.

9 A new computer is priced at £595
excluding VAT.
VAT at 17.5% must be paid on it.
Estimate the amount of VAT to be
paid.

10 A square paving slab has area
6000 cm^2. Estimate the length of a
side of the slab.

11 A circle has radius 4.3 cm.
Estimate its area.

12 Estimate the answers to these
calculations.
(a) 71×58 (b) $\sqrt{46}$

(c) $\dfrac{5987}{5.1}$ (d) 19.1^2

(e) 62.7×8316 (f) $\dfrac{5.72}{19.3}$

(g) $\dfrac{32}{49.4}$ (h) 8152×37

(i) $\dfrac{935 \times 41}{8.5}$ (j) $\dfrac{673 \times 0.76}{3.6 \times 2.38}$

EXERCISE 38.3H

Give your answers to these questions as simply as possible.

Leave π in your answers where appropriate.

1 (a) $2 \times 6 \times \pi$ (b) $\pi \times 7^2$
   (c) $\pi \times 12^2$ (d) $2 \times 3.8 \times \pi$
   (e) $\pi \times 11^2$

2 (a) $14\pi + 5\pi$
   (b) $\pi \times 3^2 + \pi \times 6^2$
   (c) $\pi \times 8^2 - \pi \times 4^2$
   (d) $3 \times 42\pi$
   (e) $\dfrac{36\pi}{4\pi}$

3 Find the circumference of a circle with radius 15 cm.

4 The areas of two circles are in the ratio $36\pi : 16\pi$. Simplify this ratio.

5 A circular piece of card has radius 12 cm. A square piece is removed, with side 5 cm.
   Find the area that is left.

EXERCISE 38.4H

1 Work out these.
   (a) 0.06×600 (b) 0.03×0.3
   (c) 0.9×0.04 (d) $(0.05)^2$
   (e) $(0.3)^2$ (f) 500×800
   (g) 30×5000 (h) 5.1×300
   (i) 20.3×2000 (j) 1.82×5000

2 Work out these.
   (a) $300 \div 20$ (b) $60 \div 2000$
   (c) $3.6 \div 20$ (d) $1.4 \div 0.2$
   (e) $2.4 \div 3000$ (f) $2.4 \div 0.03$
   (g) $0.08 \div 0.004$ (h) $5 \div 0.02$
   (i) $400 \div 0.08$ (j) $60 \div 0.15$

3 Given that $4.5 \times 16.8 = 75.6$, work out these.
   (a) 45×1680 (b) $75.6 \div 168$
   (c) $7560 \div 45$ (d) 0.168×0.045
   (e) $756 \div 0.168$

4 Given that $702 \div 39 = 18$, work out these.
   (a) $70\,200 \div 39$ (b) $70.2 \div 3.9$
   (c) 180×39 (d) $7.02 \div 18$
   (e) 1.8×3.9

5 Given that $348 \times 216 = 75\,168$, work out these.
   (a) $751\,680 \div 216$
   (b) $34\,800 \times 2160$
   (c) 3.48×21.6
   (d) $751.68 \div 34.8$
   (e) 0.348×2160

EXERCISE 39.1H

1 Pearl is a child minder. She charges £3.50 an hour.
She looks after Mrs Khan's child for 6 hours. How much does she charge?

2 It costs £60 plus £1 a mile to hire a coach.
  (a) How much does it cost to hire a coach to go
    (i) 80 miles?
    (ii) 150 miles?
  (b) Write a formula for the cost, £C, of hiring a coach to go *n* miles.

3 The cost of booking a room for a meeting is £80 plus £20 an hour.
  (a) How much does it cost to hire the room for
    (i) 5 hours?
    (ii) 8 hours?
  (b) Write a formula for the cost, £C, of hiring the room for *h* hours.

4 The perimeter of a rectangle is twice the length plus twice the width.
  (a) What is the perimeter of a rectangle with length 5 cm and width 3.5 cm?
  (b) Write a formula for the perimeter, *P*, of a rectangle with length *x* and width *y*.

5 To find the volume of a pyramid, multiply the area of the base by the height and divide by 3.
  (a) What is the volume of a pyramid with base area 12 cm^2 and height 7 cm?
  (b) Write a formula for the volume, *V*, of a pyramid with base area *A* and height *h*

6 To find the time it takes to type a document, divide the number of words in the document by the number of words typed per minute.
  (a) How long does it take Liz to type a document 560 words long if she typed 80 words a minute?
  (b) Write a formula for the time, *T*, to type a document *w* words long if the typist types *r* words a minute.

7 The time, *t*, for a journey is the distance, *d*, divided by the speed, *s*.
  (a) Write a formula for this.
  (b) Steve travelled 175 miles at a speed of 50 mph. How long did the journey take?

8 The circumference of a circle is given by the formula:
    $C = \pi \times D$, where *D* is the diameter of the circle.

  Find the circumference of a circle with diameter 8.5 cm. Use $\pi = 3.14$.

9 At Carterknowle toddlers group the charge is £1 per carer and 50p for every toddler they bring.
   (a) Tracey brings three toddlers to the group.
   How much does she pay?
   (b) Fran brings n toddlers to the group. Write down an equation for the amount, £A, she has to pay.

10 (a) For the formula $A = b - c$, find A when $b = 6$ and $c = 3.5$.
   (b) For the formula $B = 2a - b$, find B when $a = 6$ and $b = 5$.
   (c) For the formula $C = 2a - b + 3c$, find C when $a = 3.5$, $b = 2.6$ and $c = 1.2$.
   (d) For the formula $D = 3b^2$, find D when $b = 2$.
   (e) For the formula $E = ab - cd$, find E when $a = 12.5$, $b = 6$, $c = 3.5$ and $d = 8$.
   (f) For the formula $F = \dfrac{a - b}{5}$, find F when $a = 6$ and $b = 3.5$.

EXERCISE 39.2H

1 Rearrange each of these formulae to make the letter in brackets the subject.
   (a) $a = b + c$ (b)
   (b) $a = 3x - y$ (x)
   (c) $a = b + ct$ (t)
   (d) $F = 2(q + p)$ (q)
   (e) $x = 2y - 3z$ (y)
   (f) $P = \dfrac{3 + 4n}{5}$ (n)

2 The formula for the circumference of a circle is $C = \pi d$.
   Rearrange the formula to make d the subject.

3 Rearrange the formula $A = \dfrac{3ab}{2n}$
   to make
   (a) a the subject.
   (b) n the subject.

4 The formula for finding the perimeter of a rectangle is $P = 2(a + b)$, where P is the perimeter, a is the length and b is the width of the rectangle.
   Rearrange the formula to make a the subject.

5 The formula $y = mx + c$ is the equation of a straight line.
   Rearrange it to find m in terms of x, y and c.

6 The surface area of a sphere is given by the formula $A = 4\pi r^2$
   Rearrange the formula to make r the subject.

7 The formula for the volume of this prism is $V = \dfrac{\pi r^2 h}{4}$.

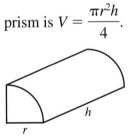

   (a) Find V when $r = 2.5$ and $h = 7$.
   (b) (i) Rearrange the formula to make r the subject.
       (ii) Find r when $V = 100$ and $h = 10$.

8 The formula for the surface area of a closed cylinder is $S = 2\pi r(r + h)$.
   Rearrange the formula to make h the subject.

EXERCISE 39.3H

1 Calculate the value of $x^3 + x$ when
   (a) $x = 2$.
   (b) $x = 3$.
   (c) $x = 2.5$.

2 (a) Calculate the value of $x^3 - x$ when
     (i) $x = 4$.   **(ii)** $x = 5$.
     (iii) $x = 4.6$.   **(iv)** $x = 4.7$.
     (v) $x = 4.65$.
   (b) Using your answers to part **(a)**,
     give the solution of $x^3 - x = 94$,
     to 1 decimal place.

Use trial and improvement for questions **3** to **10**. Show your trials.

3 Find a solution, between $x = 2$ and $x = 3$, to the equation $x^3 = 11$. Give your answer correct to 1 decimal place.

4 (a) Show that a solution to the equation $x^3 + 3x = 30$ lies between $x = 2$ and $x = 3$.
   (b) Find the solution correct to 1 decimal place.

5 (a) Show that a solution to the equation $x^3 - 2x = 70$ lies between $x = 4$ and $x = 5$.
   (b) Find the solution correct to 1 decimal place.

6 Find a solution to the equation $x^3 + 4x = 100$. Give your answer correct to 1 decimal place.

7 Find a solution to the equation $x^3 + x = 60$. Give your answer correct to 2 decimal places.

8 Find a solution to the equation $x^3 - x^2 = 40$. Give your answer correct to 2 decimal places.

9 A number, x, added to the square of that number is equal to 1000.
   (a) Write this as a formula.
   (b) Find the number correct to 1 decimal place.

10 The cube of a number minus the number is equal to 600. Find the number correct to 2 decimal places.

EXERCISE 40.1H

1 Draw a pair of axes and label them −2 to 4 for *x* and *y*.

  (a) Draw a triangle with vertices at (1, 1), (1, 3) and (0, 3). Label it A.

  (b) Reflect triangle A in the line *x* = 2. Label it B.

  (c) Reflect triangle A in the line *y* = *x*. Label it C.

  (d) Reflect triangle A in the line *y* = 2. Label it D.

2 Draw a pair of axes and label them −3 to 3 for *x* and *y*.

  (a) Draw a triangle with vertices at (−1, 1), (−1, 3) and (−2, 3). Label it A.

  (b) Reflect triangle A in the line $x = \frac{1}{2}$. Label it B.

  (c) Reflect triangle A in the line *y* = *x*. Label it C.

  (d) Reflect triangle A in the line *y* = −*x*. Label it D.

3 For each part
  • copy the diagram, making it larger if you wish.
  • reflect the shape in the mirror line.

  (a)
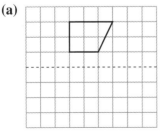

(b)
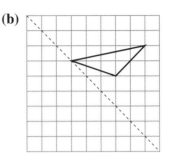

(c)
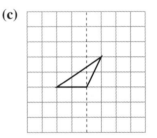

4 Describe fully the single transformation that maps
  (a) shape A on to shape B.
  (b) shape A on to shape C.
  (c) shape B on to shape D.

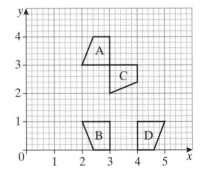

5 Describe fully the single transformation that maps
   (a) triangle A on to triangle B.
   (b) triangle A on to triangle C.
   (c) triangle E on to triangle F.

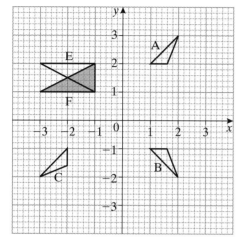

EXERCISE 40.2H

1 Copy the diagram.
   (a) Rotate trapezium A through 180° about the origin. Label it B.
   (b) Rotate trapezium A through 90° clockwise about the point (0, 1). Label it C.
   (c) Rotate trapezium A through 90° anticlockwise about the point (−1, 1). Label it D.

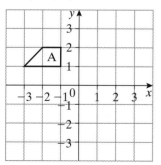

2 Copy the diagram.
   (a) Rotate flag A through 90° clockwise about the origin. Label it B.
   (b) Rotate flag A through 90° anticlockwise about the point (1, −1). Label it C.
   (c) Rotate flag A through 180° about the point (0, −1) Label it D.

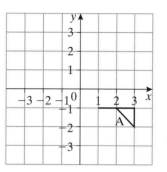

3 Draw a pair of axes and label them −4 to 4 for *x* and *y*.
   (a) Draw a triangle with vertices (1, 1), (2, 1) and (2, 3). Label it A.
   (b) Rotate triangle A through 90° anticlockwise about the origin. Label it B.
   (c) Rotate triangle A through 180° about the point (2, 1). Label it C.
   (d) Rotate triangle A through 90° clockwise about the point (−2, 1). Label it D.

4 Copy the diagram. Rotate the triangle through 180° about the point C.

5 Copy the diagram. Rotate the shape through 90° clockwise about the point O.

6 Copy the diagram. Rotate the flag through 150° clockwise about the point A.

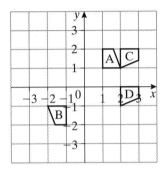

A

7 Describe fully the single transformation that maps
(a) trapezium A on to trapezium B.
(b) trapezium A on to trapezium C.
(c) trapezium A on to trapezium D.

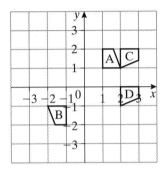

8 Describe fully the transformation that maps
(a) flag A on to flag B.
(b) flag A on to flag C.
(c) flag A on to flag D.

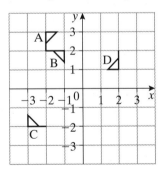

9 Describe fully the single transformation that maps triangle A on to triangle B.

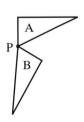

10 Describe fully the single transformation that maps
(a) triangle A on to triangle B.
(b) triangle A on to triangle C.
(c) triangle A on to triangle D.
(d) triangle A on to triangle E.
(e) triangle A on to triangle F.
Hint: Some of these transformations are reflections.

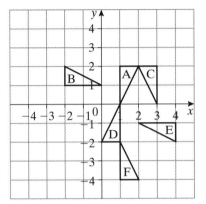

EXERCISE 40.3H

1 Draw a pair of axes and label them −2 to 6 for x and y.
(a) Draw a triangle with vertices at (1, 1), (1, 2), and (4, 1). Label it A.
(b) Translate triangle A by vector $\begin{pmatrix} 1 \\ 3 \end{pmatrix}$. Label it B.
(c) Translate triangle A by vector $\begin{pmatrix} -3 \\ 4 \end{pmatrix}$. Label it C.
(d) Translate triangle A by vector $\begin{pmatrix} -2 \\ -3 \end{pmatrix}$. Label it D.

2 Draw a pair of axes and label them −3 to 5 for x and y.
(a) Draw a triangle with vertices at (2, 1), (2, 3) and (3, 1). Label it A.

(b) Translate triangle A by vector $\begin{pmatrix} 2 \\ 1 \end{pmatrix}$.
Label it B.

(c) Translate triangle A by vector $\begin{pmatrix} -5 \\ -3 \end{pmatrix}$.
Label it C.

(d) Translate triangle A by vector $\begin{pmatrix} 2 \\ -4 \end{pmatrix}$.
Label it D.

3 Describe the single transformation that maps
(a) triangle A on to triangle B.
(b) triangle A on to triangle C.
(c) triangle A on to triangle D.
(d) triangle B on to triangle D.

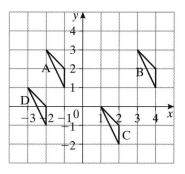

4 Describe the single transformation that maps
(a) shape A on to shape B.
(b) shape A on to shape C.
(c) shape A on to shape D.
(d) shape D on to shape E.
(e) shape A on to shape F.
(f) shape E on to shape G.
(g) shape B on to shape H.
(h) shape H on to shape F.
Hint: Not all the transformations are translations.

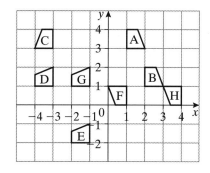

EXERCISE 40.4H

1 Draw a pair of axes and label them 0 to 6 for both x and y.
(a) Draw a triangle with vertices at (0, 6), (3, 6) and (3, 3). Label it A.
(b) Enlarge triangle A by scale factor $\frac{1}{3}$, with the origin as the centre of enlargement. Label it B.
(c) Describe fully the single transformation that maps trapezium B on to trapezium A.

2 Draw a pair of axes and label them 0 to 6 for both x and y.
(a) Draw a triangle with vertices at (5, 2), (5, 6) and (3, 6). Label it A.
(b) Enlarge triangle A by scale factor $\frac{1}{2}$, with centre of enlargement (3, 2). Label it B.
(c) Describe fully the single transformation that maps triangle B on to triangle A.

3 Draw a pair of axes and label them 0 to 8 for both x and y.
(a) Draw a triangle with vertices at (2, 1), (2, 3), (3, 2). Label it A.
(b) Enlarge triangle A by scale factor $2\frac{1}{2}$, with the origin as the centre of enlargement. Label it B.

(c) Describe fully the single transformation that maps triangle B on to triangle A.

4 Draw a pair of axes and label them 0 to 7 for both x and y.
   (a) Draw a trapezium with vertices at (1, 2), (1, 3), (2, 3) and (3, 2). Label it A.
   (b) Enlarge triangle A by scale factor 3, with centre of enlargement (1, 2). Label it B.
   (c) Describe fully the single transformation that maps triangle B on to triangle A.

5 Describe fully the single transformation that maps
   (a) triangle A on to triangle B.
   (b) triangle B on to triangle A.
   (c) triangle A on to triangle C.
   (d) triangle C on to triangle A.

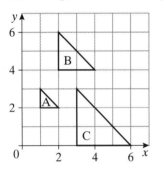

Hint: in questions **6**, **7** and **8**, not all the transformations are enlargements.

6 Describe fully the single transformation that maps
   (a) flag A on to flag B.
   (b) flag B on to flag C.
   (c) flag B on to flag D.
   (d) flag B on to flag E.
   (e) flag F on to flag G.

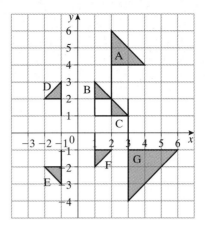

7 Draw a pair of axes and label them -4 to 4 for x and y.
   (a) Draw a triangle with vertices at (2, 1), (2, 2) and (4, 2). Label it A.
   (b) Reflect triangle A in the line $y = 0$. Label it B.
   (c) Reflect triangle A in the line $x = 1$. Label it C.
   (d) Rotate triangle B by 90° anticlockwise about the origin. Label it D.
   (e) Enlarge triangle A by scale factor $\frac{1}{2}$, with the origin as the centre of enlargement. Label it E.

8 Copy the diagram.
   (a) Rotate shape A through 90° clockwise about the origin. Label it B.
   (b) Rotate shape A through 180° about (2, 2). Label it C.
   (c) Enlarge shape A by scale factor $\frac{1}{2}$, with centre of enlargement $(-2, 0)$. Label it E.

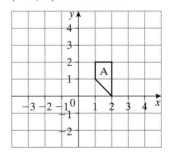

EXERCISE 41.1H

1 The probability that Stacey will go to bed late tonight is 0.2.
   What is the probability that Stacey will not go to bed late tonight?

2 The probability that I will throw a six with a dice is $\frac{1}{6}$.
   What is the probability that I will not throw a six?

3 The probability that it will snow on Christmas Day is 0.15.
   What is the probability that it will not snow on Christmas Day?

4 The probability that someone chosen at random is left-handed is $\frac{3}{10}$.
   What is the probability that they will be right-handed?

5 The probability that United will lose their next game is 0.08.
   What is the probability that United will not lose their next game?

6 The probability that Ian will eat crisps one day is $\frac{17}{31}$.
   What is the probability that he will not eat crisps?

EXERCISE 41.2H

1 A shop has brown, white and wholemeal bread for sale.
   The probability that someone will choose brown bread is 0.4 and the probability that they will choose white bread is 0.5.
   What is the probability of someone choosing wholemeal bread?

2 A football coach is choosing a striker for the next game.
   He has three players to choose from; Wayne, Michael and Alan.
   The probability that he will choose Wayne is $\frac{5}{19}$ and the probability that he will choose Michael is $\frac{7}{19}$.
   What is the probability that he will choose Alan?

3 A bag contains red, white and blue counters. Jill chooses a counter at random.
   The probability that she chooses a red counter is 0.4 and the probability that she chooses a blue counter is 0.15.
   What is the probability that she chooses a white counter?

4 Elaine goes to town by car, bus, taxi or bike.
The probability that she uses her car is $\frac{12}{31}$, the probability that she catches the bus is $\frac{2}{31}$ and the probability that she takes a taxi is $\frac{13}{31}$.
What is the probability that she rides her bike into town?

5 A biased five-sided spinner is numbered 1 to 5. The table shows the probability of obtaining some of the scores when it is spun.

Score	1	2	3	4	5
Probability	0.37	0.1	0.14		0.22

What is the probability of getting 4?

6 A cash bag contains only £20, £10 and £5 notes. One note is chosen from the bag at random.
There is a probability of $\frac{3}{4}$ that it is a £5 note and a probability of $\frac{3}{20}$ that it is a £10 note.
What is the probability that it is a £20 note?

EXERCISE 41.3H

1 The probability that United will lose their next game is 0.2.
How many games would you expect them to lose in a season of 40 games?

2 The probability that it will be rainy on any day in June is $\frac{2}{15}$.
On how many of June's 30 days would you expect it to be rainy?

3 The probability that an eighteen-year-old driver will have an accident is 0.15. There are 80 eighteen-year-old drivers in a school.
How many of them might be expected to have an accident?

4 When Phil is playing chess, the probability that he wins is $\frac{17}{20}$.
In a competition, Phil plays 10 games. How many of them might you expect him to win?

5 An ordinary six-sided dice is thrown 90 times. How many times might you expect to get
(a) a 4?   (b) an odd number?

6 A box contains twelve yellow balls, three blue balls and five green balls.
A ball is chosen at random and its colour noted.
The ball is then replaced. This is done 400 times.
How many of each colour might you expect to get?

EXERCISE 41.4H

1 Pete rolls a dice 200 times and records the number of times each score appears.

Score	1	2	3	4	5	6
Frequency	29	34	35	32	34	36

(a) Work out the relative frequency of each of the scores.
Give your answers to 2 decimal places.
(b) Do you think that Pete's dice is fair? Give a reason for your answer.

2 Rory kept a record of his favourite football team's results.

Win: 32   Draw: 11   Lose: 7

(a) Calculate the relative frequency of each of the three outcomes.

(b) Are your answers to part (a) good estimates of the probability of the outcome of their next match? Give a reason for your answer.

3 In a survey, 600 people were asked which flavour of crisps they preferred. The results are shown in the table.

Flavour	Frequency
Plain	166
Salt & Vinegar	130
Cheese & Onion	228
Other	76

(a) Work out the relative frequency for each flavour. Give your answers to 2 decimal places.

(b) Explain why it is reasonable to use these figures to estimate the probability of the flavour of crisps that the next person to be asked will prefer.

4 The owner of a petrol station notices that in one day 287 out of 340 people filling their car with petrol spent over £20.

Use these figures to estimate the probability that the next customer will spend

(a) over £20.   (b) £20 or less.

5 Jasmine made a spinner numbered 1, 2, 3, 4 and 5.
She tested the spinner to see if it was fair.
The results are shown below.

Score	1	2	3	4	5
Frequency	46	108	203	197	96

(a) Work out the relative frequency of each of the scores. Give your answers to 2 decimal places.

(b) Do you think that the spinner is fair? Give a reason for your answer.

6 A box contains yellow, green, white and blue counters.
A counter is chosen from the box and its colour noted. The counter is then replaced in the box.
The table below gives information about the colour of counter picked.

Colour	Relative frequency
Yellow	0.4
Green	0.3
White	0.225
Blue	0.075

(a) There are 80 counters altogether in the bag.
How many do you think there are of each colour?

(b) What other information is needed before you can be sure that your answers to part (a) are accurate?

EXERCISE 42.1H

1 Draw the graph of $y = 3x$ for values of x from -3 to 3.

2 Draw the graph of $y = x + 2$ for values of x from -4 to 2.

3 Draw the graph of $y = 4x + 2$ for values of x from -3 to 3.

4 Draw the graph of $y = 2x - 5$ for values of x from -1 to 5.

5 Draw the graph of $y = -2x - 4$ for values of x from -4 to 2.

EXERCISE 42.2H

1 Draw the graph of $3y = 2x + 6$ for values of x from -3 to 3.

2 Draw the graph of $2x + 5y = 10$.

3 Draw the graph of $3x + 2y = 15$.

4 Draw the graph of $2y = 5x - 8$, for values of x from -2 to 4.

5 Draw the graph of $3x + 4y = 24$.

EXERCISE 42.3H

1 Simon had a bath. The graph shows the volume (V gallons) of the water in the bath after t minutes.

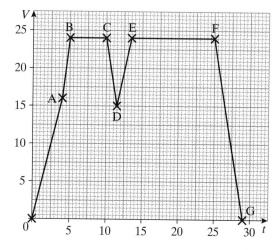

(a) How many gallons of water are in the bath at A?

(b) Simon got in the bath at B and out at F. How long was he in the bath?

(c) Between O and A, the hot tap is on. How many gallons of water per minute came from the hot tap?

(d) Between A and B, both taps are on. What is the rate of flow of both taps together?
Give your answer in gallons/minute.

(e) Describe what happened between C and E.

(f) At what rate did the bath empty?
Give your answer in gallons/minute.

2 A printer's charge for printing programmes is worked out as follows.

A fixed charge of £*a*
+
x pence per programme for the first 1000 programmes
+
80 pence per programme for each programme over 1000

The graph below shows the total charge for printing up to 1000 programmes.

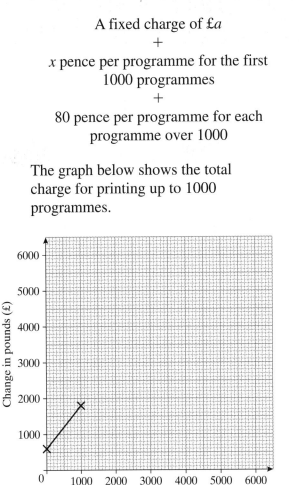

(a) What is the fixed charge, £*a*?
(b) Calculate *x*, the charge per programme for the first 1000 programmes.
(c) Copy the graph and add a line segment to show the charges for 1000 to 6000 programmes.
(d) What is the total charge for 3500 programmes?
(e) What is the average cost per programme for 3500 programmes?

3 The graph shows a train journey.

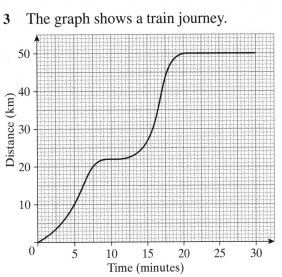

(a) How long did the train journey take?
(b) How far was the train journey?
(c) How far from the start was the first station?
(d) How long did the train stop at the first station?
(e) When was the train travelling fastest?

4 Water is poured into each of these vessels at a constant rate until they are full.

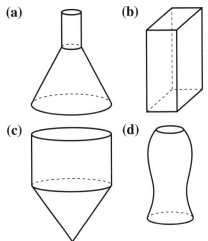

These graphs show depth of water (*d*) against time (*t*).

Choose the most suitable graph for each vessel.

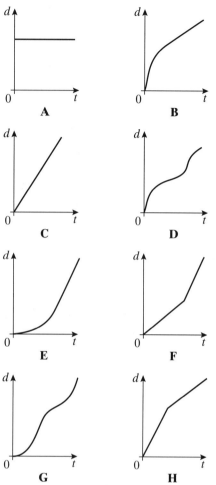

5   A mobile phone company offers its customers the choice of two price plans.

	Plan A	Plan B
Monthly subscription	£10.00	£s
Free talk time per month	60 minutes	100 minutes
Cost per minute over the free talk time	a pence	35 pence

The graph shows the charges for Plan A and the charges for Plan B up to 100 minutes.

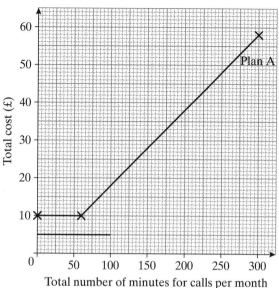

(a) Find the monthly subscription for Plan B (£s).
(b) Shamir uses price plan B. How much does it cost him if he uses the phone for 250 minutes per month?
(c) Copy the graph and add a line to show the cost for Plan B for 100 to 250 minutes.
(d) For how many minutes is the cost the same in both price plans?
(e) Which price plan is the cheaper when the time for calls is 220 minutes? By how much?

6 The table shows the cost of sending parcels.
The graph shows the information in the first row in the table.

Maximum weight	Cost
10 kg	£13.85
11 kg	£14.60
12 kg	£15.35
13 kg	£16.10
14 kg	£16.85
15 kg	£17.60

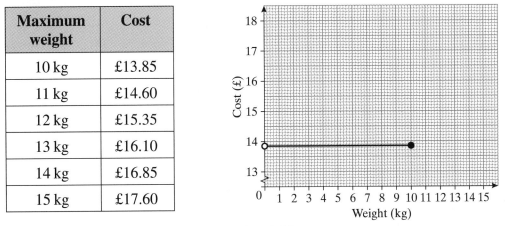

(a) What is the cost of sending a parcel weighing
   (i) 9.6 kg?   (ii) 10 kg?   (iii) 10.1 kg?

(b) (i) What is the meaning of the dot at the right of the line?
    (ii) What is the meaning of the circle at the left of the line?

(c) Copy the graph and add lines to show the cost for parcels weighing up to 15 kg.

(d) Hazel posted one parcel weighing 8.4 kg and another weighing 12.8 kg. What was the total cost?

7 John and Hywel are brothers living in the same house. The graph shows John's cycling trip from their home. He cycles for an hour, stops for a rest, then continues his journey.

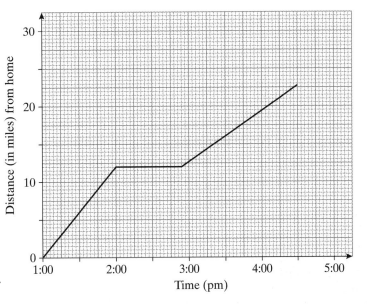

(a) How far did John travel in the first 45 minutes?

(b) For how many minutes did John rest?

(c) Hywel sets out from their home at 3 pm. and travels by car at an average speed of 25 m.p.h. on the same route as John.
Draw his journey on the graph paper.

(d) Write down the time at which Hywel passes John.

WJEC Summer 2005

EXERCISE 42.4H

1 Which of these functions are quadratic? For each of the functions that is quadratic, state whether the graph is ∪-shaped or ∩-shaped.

(a) $y = x^2 + 7$ (b) $y = 2x^3 + x^2 - 4$ (c) $y = x^2 + x - 7$ (d) $y = x(6 - x)$

(e) $y = \dfrac{5}{x^2}$ (f) $y = x(x^2 + 1)$ (g) $y = 2x(x + 2)$ (h) $y = 5 + 3x - x^2$

2 (a) Copy and complete the table of values for $y = 2x^2$.

x	-3	-2	-1	0	1	2	3
x^2	9					4	
$y = 2x^2$	18					8	

(b) Plot the graph of $y = 2x^2$.
Use a scale of 2 cm to 1 unit on the x-axis and 1 cm to 1 unit on the y-axis.

(c) Use your graph to
   (i) find the value of y when $x = -1.8$.   (ii) solve $2x^2 = 12$.

3 (a) Copy and complete the table of values for $y = x^2 + x$.

x	-4	-3	-2	-1	0	1	2	3
x^2			4					9
$y = x^2 + x$								12

(b) Plot the graph of $y = x^2 + x$.
Use a scale of 2 cm to 1 unit on the x-axis and 1 cm to 1 unit on the y-axis.

(c) Use your graph to
   (i) find the value of y when $x = 1.6$.   (ii) solve $x^2 + x = 8$.

4 (a) Copy and complete the table of values for $y = x^2 - x + 2$.

x	-3	-2	-1	0	1	2	3	4
x^2		4						16
$-x$		2						-4
2		2						2
$y = x^2 - x + 2$		8						14

(b) Plot the graph of $y = x^2 - x + 2$.
Use a scale of 2 cm to 1 unit on the x-axis and 1 cm to 1 unit on the y-axis.

(c) Use your graph to
   (i) find the value of y when $x = 0.7$ (ii) solve $x^2 - x + 2 = 6$.

5   (a) Copy and complete the table of values for $y = x^2 + 2x - 5$.

x	−5	−4	−3	−2	−1	0	1	2	3
x^2				4					9
$2x$				−4					6
−5				−5					−5
$y = x^2 + 2x - 5$				−5					10

(b) Plot the graph of $y = x^2 + 2x - 5$.
Use a scale of 2 cm to 1 unit on the x-axis and 1 cm to 1 unit on the y-axis.
(c) Use your graph to
   (i)   find the value of y when $x = -1.4$.   (ii)   solve $x^2 + 2x - 5 = 0$.

6   (a) Copy and complete the table of values for $y = 8 - x^2$.

x	−3	−2	−1	0	1	2	3
8				8			8
$-x^2$				0			−9
$y = 8 - x^2$				8			−1

(b) Plot the graph of $y = 8 - x^2$.
Use a scale of 2 cm to 1 unit on the x-axis and 1 cm to 1 unit on the y-axis.
(c) Use your graph to
   (i)   find the value of y when $x = 0.5$.   (ii)   solve $8 - x^2 = -2$.

7   (a) Copy and complete the table of values for $y = (x - 2)(x + 1)$.

x	−3	−2	−1	0	1	2	3	4
$x - 2$		−4					1	
$x + 1$		−1					4	
$y = (x - 2)(x + 1)$		4					4	

(b) Plot the graph of $y = (x - 2)(x + 1)$.
Use a scale of 2 cm to 1 unit on the x-axis and 1 cm to 1 unit on the y-axis.
(c) Use your graph to
   (i)   find the minimum value of y.   (ii)   solve $(x - 2)(x + 1) = 2.5$.

8   (a) Make a table of values for $y = x^2 - 3x + 2$. Choose values of x from −2 to 5.
(b) Plot the graph of $y = x^2 - 3x + 2$.
Use a scale of 2 cm to 1 unit on the x-axis and 1 cm to 1 unit on the y-axis.
(c) Use your graph to solve
   (i) $x^2 - 3x + 2 = 1$.   (ii) $x^2 - 3x + 2 = 10$.

EXERCISE 43.1H

1 Change these units.
   (a) 25 cm to mm
   (b) 24 m to cm
   (c) 1.36 cm to mm
   (d) 15.1 cm to mm
   (e) 0.235 m to mm

2 Change these units.
   (a) 2 m² to cm²
   (b) 3 cm² to mm²
   (c) 1.12 m² to cm²
   (d) 0.05 cm² to mm²
   (e) 2 m² to mm²

3 Change these units.
   (a) 8000 mm² to cm²
   (b) 84 000 mm² to cm²
   (c) 2 000 000 cm² to m²
   (d) 18 000 000 cm² to m²
   (e) 64 000 cm² to m²

4 Change these units.
   (a) 32 cm³ to mm³
   (b) 24 m³ to cm³
   (c) 5.2 cm³ to mm³
   (d) 0.42 m³ to cm³
   (e) 0.02 cm³ to mm³

5 Change these units.
   (a) 5 200 000 cm³ to m³
   (b) 270 000 mm³ to cm³
   (c) 210 cm³ to m³
   (d) 8.4 m³ to mm³
   (e) 170 mm³ to cm³

6 Change these units.
   (a) 36 litres to cm³
   (b) 6300 ml to litres
   (c) 1.4 litres to ml
   (d) 61 ml to litres
   (e) 5400 cm³ to litres.

EXERCISE 43.2H

1 Copy and complete each of these statements.
   (a) A length given as 4.3 cm, to 1 decimal place, is between .... cm and .... cm.
   (b) A capacity given as 463 ml, to the nearest millilitre, is between .... ml and .... ml
   (c) A time given as 10.5 seconds, to the nearest tenth of a second, is between .... seconds and .... seconds.
   (d) A mass given as 78 kg, to the nearest kilogram, is between .... kg and .... kg.
   (e) An area given as 5.5 m², to 1 decimal place, is between .... m² and .... m².

2 The number of people attending a football match was given as 24 000 to the nearest thousand.
What was the least number of people that could have been at the match?

3 Kerry measures her height as 142 cm to the nearest centimetre.
Write down the two values between which her height must lie.

4 The height of a desk is stated as 75.0 cm to 1 decimal place.
Write down the two values between which its height must lie.

5 Rashid measures the thickness of a piece of plywood as 7.83 mm, to 2 decimal places.
Write down the smallest and greatest thickness it could be.

6 The sides of this triangle are given in centimetres, correct to 1 decimal place.

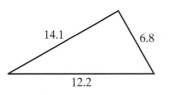

(a) Write down the shortest and longest possible length of each side.
(b) Write down the shortest and longest possible length of the perimeter.

7 John has two pieces of string.
He measures them as 125 mm and 182 mm, to the nearest millimetre.
He puts the two pieces end to end.
What is the shortest and longest that their combined lengths can be?

8 Mel and Mary both buy some apples.
Mel buys 3.5 kg and Mary buys 4.2 kg.
Both weights are correct to the nearest tenth of a kilogram.
(a) What is the smallest possible difference between the amounts they have bought?
(b) What is the largest possible difference between the amounts they have bought?

EXERCISE 43.3H

1 Rewrite each of these statements using sensible values for the measurements.
(a) My mass is 78.32 kg
(b) It takes Katriona 16 minutes and 15.6 seconds to walk to school.
(c) The distance to London from Sheffield is 161.64 miles.
(d) The length of our classroom is 5 metres 14 cm 3 mm.
(e) My water jug hold 3.02 litres.

2 Give your answer to each of these questions to a sensible degree of accuracy.
(a) Estimate the length of this line.

(b) Estimate the size of this angle.

(c) A rectangle is 2.3 cm long and 4.5 cm wide. Find the area of the rectangle.

(d) The diameter of a circle is 8 cm. Work out the circumference.

(e) The volume of a cube is 7 cm³. Find the length of an edge.

(f) An angle in a pie chart is found by working out $\frac{4}{7} \times 360°$. Find the angle.

(g) A coach travels 73 miles at an average speed of 33 mph. How long does it take?

(h) Six friends share £14 between them. How much does each one get?

EXERCISE 43.4H

1 A boat travels 24 km in 3 hours. Calculate its average speed.

2 A car covers 197 miles on a motorway in 3 hours. Calculate the average speed. Give your answer to 1 decimal place.

3 Anne walks at an average speed of 3.5 km/h for 2 hours 30 minutes. How far does she walk?

4 How long will it take a boat sailing at 12 km/h to travel 64 km?

5 The density of a rock is 9.3 g/cm³. Its volume is 60 cm³. What is its mass?

6 Calculate the density of a piece of metal with mass 300 g and volume 84 cm³. Give your answer to a suitable degree of accuracy.

7 A man walks 10 km in 2 hours 15 minutes. What is his average speed in km/h? Give your answer to a suitable degree of accuracy.

8 Calculate the mass of a stone of volume 46 cm³ and density 7.6 g/cm³.

9 Copper has a density of 8.9 g/cm³. Calculate the volume of a block of copper of mass 38 g. Give your answer to a suitable degree of accuracy.

10 What is the density of gas if a mass of 32 kg occupies a volume of 25 m³? Give your answer to a suitable degree of accuracy.

11 A small town in America has a population of 235 and covers an area of 35 km². Find the population density (number of people per square kilometre) of the town.

12 A coach left at 0910 and travelled 72 km in 90 minutes, arriving at Carter castle.

(a) Calculate its average speed.

The coach stopped at the castle for $3\frac{1}{2}$ hours and then travelled back at an average speed of 55 km/h.

(b) What time did it arrive back? Give your answer to the nearest minute.

EXERCISE 44.1H

1 State whether the following are primary data or secondary data.
  (a) Weighing packets of sweets
  (b) Using bus timetables
  (c) Looking up holiday prices on the internet
  (d) A GP entering data for a new patient on his records after seeing the patient

2 Lisa is doing a survey and has written this question.

> **What colour is your hair?**
> Black ☐   Brown ☐   Blonde ☐

  (a) Give a reason why this question is unsuitable.
  (b) Write a better version.

3 Steve is doing a survey about his local sports facilities.
  Here is one of his questions.

> **How much do you enjoy doing sport?**
>   1   2   3   4   5

  (a) Give a reason why this question is unsuitable.
  (b) Write a better version.

4 Mia is doing a survey about school lunches.
  She gives out questionnaires to the first 30 people in the queue for lunch.
  (a) Why is this likely to give a biased sample?
  (b) Describe a better method of obtaining a sample for her survey.

5 Here is one of Mia's questions.

> **Don't you agree that we don't have enough salads on the menu?**

  (a) Give a reason why this question is unsuitable.
  (b) Write a better version.

6 A survey is to be done about school students' earnings and pocket money. Write five suitable questions which should be included in such a survey.

EXERCISE 45.1H

1 Look at this sequence of circles.
The first four patterns in the sequence
have been drawn.

   (a) How many circles are there in the
100th pattern?
   (b) Describe the position to term rule
for this sequence.

2 Look at this sequence of matchstick
patterns.

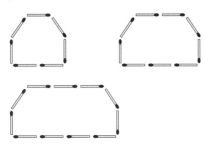

   (a) Complete this table.

Pattern number	1	2	3	4	5
Number of matchsticks					

   (b) What patterns have you noticed in
the numbers?
   (c) Find the number of matchsticks in
the 50th pattern.

3 Here is a sequence of star patterns.

   (a) Draw the next pattern in the
sequence.
   (b) Without drawing the pattern, find the
number of stars in the 8th pattern.
Explain how you found your answer.

4 The numbers in a sequence are given
by this rule:

   Multiply the position number by 7,
then subtract 10.

   (a) Show that the first term of the
sequence is -3.
   (b) Find the next four terms in the
sequence.

5 Find the first four terms of the
sequences with these nth terms.
   (a) $10n$ (b) $8n + 2$

6 Find the first five terms of the
sequences with these nth terms.
   (a) n^2 (b) $2n^2$ (c) $5n^2$

7 The first term of a sequence is 3.
The general rule for the sequence is
multiply a term by 3 to get to the next
term.
Write down the first five terms of the
sequence.

8 For a sequence, $T_1 = 12$ and
$T_{n+1} = T_n - 5$.
Write down the first four terms of this
sequence.

9 Draw suitable patterns to represent this
sequence.

$$1, 4, 7, 10, ...$$

10 Draw suitable patterns to represent this
sequence.

$$1 \times 1, 3 \times 3, 5 \times 5, 7 \times 7, ...$$

EXERCISE 45.2H

1 Find the nth term for each of these
sequences.
(a) 10, 13, 16, 19, 22, ...
(b) 0, 1, 2, 3, 4, ...
(c) $-3, -1, 1, 3, 5, ...$

2 Find the nth term for each of these
sequences.
(a) 25, 20, 15, 10, 5, ...
(b) $4, 2, 0, -2, -4, ...$
(c) $3, 2, 1, 0, -1, ...$

3 Which of these sequences are linear?
Find the next two terms of each of the
sequences that are linear.
(a) 2, 5, 10, 17, ...
(b) 2, 5, 8, 11, ...
(c) 1, 3, 6, 10, ...
(d) $12, 8, 4, 0, -4, ...$

4 (a) Write the first five terms of the
sequence with nth term $100n$.
(b) Compare your answers with this
sequence.
99, 199, 299, 399, ...
Write down the nth term of this
sequence.

5 A mail-order shirt company charges
£25 per shirt, plus an overall delivery
charge of £3.
(a) Complete the table.

Number of shirts	1	2	3
Cost in £			

(b) Write an expression for the cost, in
pounds, of n shirts.
(c) Paul pays £128 for shirts. How
many does he buy?

6 (a) Write down the first five terms of
the sequence with nth term n^2.
(b) Compare your answers with this
sequence.
0, 3, 8, 15, 24, ...
Write the nth term of this sequence.

7 The nth triangle number is $\dfrac{n(n + 1)}{2}$.

Find the 60th triangle number.

8 The nth term of a sequence is 2^n.
(a) Write down the first five terms of
this sequence.
(b) Describe the sequence.

9 (a) Write down the first five cube
numbers.
(b) Compare this sequence with the
cube numbers.
3, 10, 29, 66, 127, ...
Use what you notice to write down
the nth term of this sequence.
(c) Find the 10th term of this sequence.

10 Find the nth term of each of the
following sequences:
(a) $1 \times 2, 2 \times 3, 3 \times 4, 4 \times 5, ...$
(b) $1 \times 3, 2 \times 5, 3 \times 7, 4 \times 9, ...$
(c) $2 \times 1, 4 \times 3, 6 \times 5, 8 \times 7, ...$

EXERCISE 46.1H

1 Two points, A and B, are 7 cm apart.
   Draw the locus of points that are
   equidistant from A and B.

2 A badger will never go further than
   3 miles from its home.
   Draw a scale diagram to show the
   regions where the badger might go
   looking for food.

3 Draw an equilateral triangle, ABC, of
   side 6 cm.
   Shade the region of points inside the
   triangle which are nearer to AB than
   to AC.

4 A rectangular garden measures 8 m by
   6 m. A fence is built from F, at a right
   angle across the garden.
   Draw a scale diagram and construct
   the line of the fence.

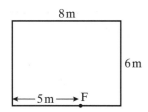

5 Draw a square, ABCD, of side 5 cm.
   Draw the locus of points, inside the
   square, which are more than 3 cm
   from A.

6 Zeke is walking across a field.
   He notices a bull starting to chase after
   him. He runs the shortest distance to
   the hedge.
   Copy the diagram and draw the path
   that Zeke should run.

• Zeke

7 Draw a rectangle, PQRS, with sides
   PQ = 7 cm and QR = 5 cm.
   Shade the region of points that are
   closer to P than to Q.

8 Draw an angle of 80°.
   Draw the bisector of the angle.

9 An office is a rectangle measuring
   16 m by 12 m. There are two
   electricity points in the office at
   opposite corners of the room. The
   vacuum cleaner has a wire 10 m long.
   Make a scale drawing to show how
   much of the room can be cleaned.

10 Make another scale drawing of the
   office in question **9**.
   Shade the locus of points which are
   equidistant from the two electricity
   points.
   Use this locus to work out the length
   of wire needed for the vacuum cleaner
   to reach everywhere in the office.

EXERCISE 46.2H

1 Draw a point and label it P.
Construct the locus of points that are
less than 4 cm from P and more than
6 cm from P.

2 A rectangular garden measures 20 m
by 12 m.
A tree is to be planted so that it is more
than 4 m from each corner of the garden.
Make a scale drawing to find the area
where the tree can be planted.

3 Two points, A and B, are 5 cm apart.
Find the region that is less than 3 cm
from A and more than 4 cm from B.

4 Make an accurate drawing of a
triangle, PQR, where PQ = 6 cm,
P = 40° and Q = 35°.
Find the point X, which is 2 cm from
R and equidistant from P and Q.

5 Two coastguard stations, A and B, are
20 km apart on a straight coastline.
The coastguard at A knows that a ship
is within 15 km of him.
The coastguard at B knows that the
same ship is within 10 km of him.
Make a scale drawing to show the
region where the ship could be.

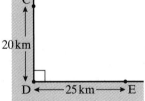

6 Two lines, each 6 cm long, join to
form a right angle.
Draw the region of points which are
less than 3 cm from these lines.

7 Two points, P and Q, are 7 cm apart.
Find the points which are the same
distance from P and Q and are also
within 5 cm of Q.

8 The diagram shows three coastguard
stations, C, D and E.
A ship is within 25 km of C and closer
to DE than DC.
Find the region where the ship could be.

9 A garden is a rectangle, ABCD, with
AB = 5 m and BC = 3 m.
A new flower bed is to be made in the
garden.
It must be more than 2 m from A and
less than 1.5 m from CD.
Make a scale drawing to show where
the flower bed could be.

10 EFG is a triangle with EF = 6 cm,
FG = 8 cm and EG = 10 cm.
Draw the perpendicular from F to EG.
Indicate the points on this line that are
more than 7 cm from G.

EXERCISE 47.1H

Work these out on your calculator without writing down the answers to the middle stages.

If the answers are not exact, give them correct to 2 decimal places.

1 $\dfrac{7.3 + 8.5}{5.7}$

2 $\dfrac{158 + 1027}{125}$

3 $\dfrac{6.7 + 19.5}{12.2 - 5.7}$

4 $\sqrt{128 - 34.6}$

5 $5.7 + \dfrac{1.89}{0.9}$

6 $(12.6 - 9.8)^2$

7 $\dfrac{8.9}{2.3 \times 5.6}$

8 $\dfrac{15.4}{2.3^2}$

9 $10.9 \times (7.2 - 5.8)$

10 $\dfrac{4.8 + 6.2}{5.2 \times 6.5}$

11 $\dfrac{7.1}{\sqrt{15.3 \times 0.6}}$

12 $\dfrac{3 - \sqrt{2.73} + 5.1}{4}$

EXERCISE 47.2H

Do not use your calculator for questions 1 to 4.

1 These calculations are all wrong. This can be spotted quickly without working them out.
For each one, give a reason why it is wrong.
(a) $15.3 \div -5.1 = 5$
(b) $8.7 \times 1.6 = 5.4375$
(c) $4.7 \times 300 = 9400$
(d) $7.5^2 = 46.25$

2 These calculations are all wrong. This can be spotted quickly without working them out.
For each one, give a reason why it is wrong.
(a) $5.400 \div 9 = 60$
(b) $-6.2 \times -0.5 = -93.1$
(c) $\sqrt{0.4} = 0.2$
(d) $8.5 \times 7.1 = 60.36$

3 Estimate the answers to each of these calculations. Show your working.
(a) 93×108
(b) 0.61^2
(c) $-19.6 + 5.2$

4 Estimate the answers to these calculations. Show your working.
(a) The cost of three DVDs at £17.99.
(b) The cost of 39 cinema tickets at £6.20.
(c) The cost of five meals at £7.99 and two drinks at £2.10.

5 Use inverse operations to check these calculations. Write down the operations you use.
(a) $19\,669.5 \div 235 = 83.7$
(b) $\sqrt{5069.44} = 71.2$
(c) $9.7 \times 12.4 = 120.28$
(d) $17.2 \times 4.6 + 68.2 = 147.32$

6 Work these out. Round your answers to 2 decimal places.

(a) $\dfrac{24.3 + 18.6}{2.8 \times 0.51}$

(b) $(13.7 + 53.1) \times (9.87 - 5.9)$

7 Work these out. Round your answers to 3 decimal places.

(a) $\dfrac{77.8}{6.4 + 83.9}$

(b) $1.06^4 \times 185$

8 Work these out. Round your answers to 2 significant figures.

(a) $\sqrt{2.5^2 + 9.0}$

(b) 640×0.078

9 (a) Use rounding to 1 significant figure to estimate the answer to each of these calculations. Show your working.

(i) 21.2^3

(ii) 189×0.31

(iii) $\sqrt{11.1^2 - 4.8^2}$

(iv) $\dfrac{51.8 + 39.2}{0.022}$

(b) Use your calculator to find the correct answer to each of the calculations in (a). Where appropriate, round your answer to a sensible degree of accuracy.

EXERCISE 47.3H

1 Write each of these times as a decimal.

(a) 8 hours 39 minutes

(b) 5 hours 21 minutes

(c) 33 minutes

(d) 3 minutes

2 Write each of these times in hours and minutes.

(a) 4.8 hours   (b) 5.85 hours

(c) 0.45 hours   (d) 0.6 hours

3 (a) A walker covers a distance of 7.2 miles in 2 hours and 24 minutes. Calculate the average speed of the walker in miles per hour.

(b) For one 25-minute section of its journey an express train travels at an average speed of 126 miles per hour. How long is this section of the journey? Give your answer in miles.

(c) In one stage of its flight a rocket travelled at a constant speed of 357 miles per hour for a distance of 23.8 miles. How long did it take the rocket to travel that distance?

4 In a four-part endurance race, a driver takes the following times to complete each part of the race.

5 hours 38 minutes
6 hours 57 minutes
5 hours 19 minutes
5 hours 46 minutes

What was the driver's total time for the race? Give your answer in hours and minutes.

5 Wayne buys some potatoes at £1.25 a kilogram and six nectarines at 37p each.
He gives the shop assistant £10 and gets £4.68 change.
What weight of potatoes did he buy?

6 Chelsea, Sally and James share the profits from their business in the ratio 4 : 3 : 2.
In 2005 the total profit was £94 500.
Calculate how much Sally received.

7 Merry followed a recipe for lemon pudding which used 350 g of flour for four people.
He made the recipe for 10 people and used a new 1.5 kg bag of flour.
How much flour did he have left?

8 Freeville has a population of 36 281 and its area is 27.4 km^2.
Calculate its population density. Give your answer to a sensible degree of accuracy.

9 Mr Brown's mobile phone bill one month showed that he had used 53 minutes of calls at 13p per minute.
His monthly rental charge for the phone was £15.30.
There was VAT at 17.5% on the whole bill.
Calculate the total bill including VAT.

10 In January 2000, the Retail Price Index (RPI) was 166.6.
In January 2001 it was 171.1.
Calculate the percentage increase in the RPI over that year.

11 In September 2003, the Average Earnings Index (AEI) was 113.9. During the next year it increased by 4.2%.
Calculate the AEI in September 2004.

12 A metal cylinder has radius 3.8 cm and height 5.7 cm.
Its density is 12 g/cm^3.
Calculate its mass.

13 A water urn is in the shape of a cuboid, with the base a square of side 30 cm.
How many litres of water does the urn contain when it is filled to a depth of 42 cm?

14 David drove 28 miles along the motorway at 70 mph, and then 10 miles at 50 mph.
(a) Calculate how long he took.
(b) Find his average speed for the whole journey.